Name: ....

"JAIN 108 MATHEMAGICS CURRICULUM For The GLOBAL SCHOOL, Book 4"

subtitled:

Workbook Level 1

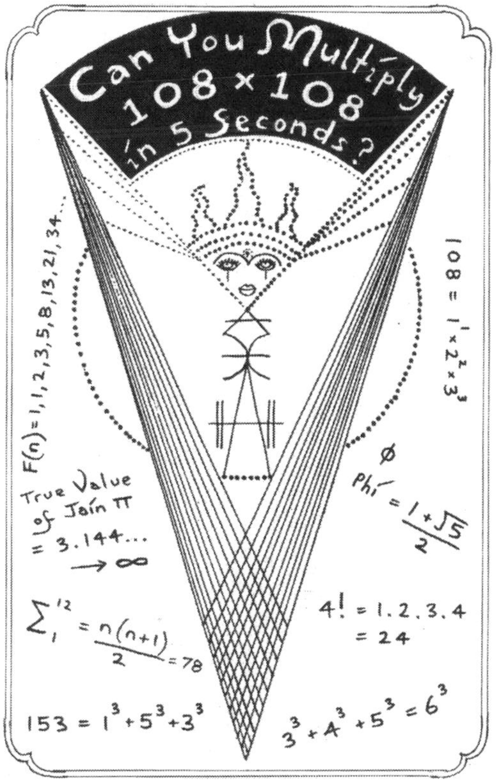

AF886305

RAPID MENTAL CALCULATION for TEENS and ADULTS

Designed to release the Inner Genius

(Script Material for the Film and OnLine Tutorials).

By **JAIN 108**

(writer, artist, **Mathematical Futurist** and EcoPreneur)

(**F**or **R**esearch **E**xpressing **E**ssential **D**ata **O**f **M**agic **S**quares)

www.jainmathemagics.com

ISBN: 978-1-925834-18-5

Level 1 Workbook
MATHEMAGICS for TEENAGERS

2011

CERTIFIED CERTIFICATE COURSE
10 Wednesday Afternoons from 4pm to 6pm

CONDUCTED by **JAIN**

CURRICULUM NOTES

Presented as Worksheets
As a companion Manual for the 5 DVD Set
from DAY 1 to DAY 10

LIST of **WORKSHEETS** to be GIVEN to the STUDENTS
Nb: This booklet is also a GUIDELINE or MANUAL for
TEACHER TRAINING SEMINARS

This COURSE REPEATS EACH SEASON or TERM
in any village, school, institution or country of the world

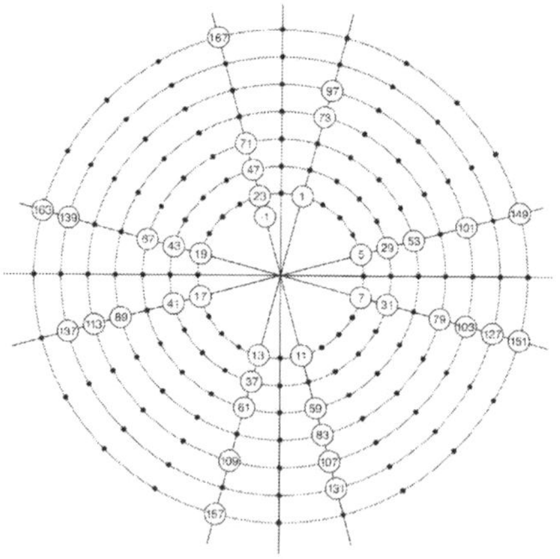

There is also a **Level 2 MATHEMAGICS for TEENAGERS** comprised of another 10 Lectures. Level 1 is therefore an Introduction and Level 2 is little more advanced.

JAIN MATHEMAGICS CURRICULUM
for the GLOBAL SCHOOL

formerly known as: The VEDIC MATHEMATICS CURRICULUM for the GLOBAL SCHOOL

Part 2 MULTIPLICATION JAIN 2005

Developing the Inner Mental Screen

Jain: International Lecturer on Vedic Mathematics and Sacred Geometry)

CONTENTS

page 70

DAY 6: ~ **Base 20**, **Base 30**,
Sub-Sutra: "**The Remainder Remains The Same**"
~ **Squaring of the Numbers from 50 to 60** eg: 52^2
~ **Base 90**: By The Deficiency, Under the Base 100 to solve 98x97 and using Sub-Sutra: "**Lessen By The Deficiency**"
~ **Base 100**: **By The Excess**, Over the Base eg: 108x109
~ **Base 1,000** and higher eg: 1011x1023

page 96

DAY 7: ~ **Cognate Numbers** eg: 22x28
~ **Sutra: "By One More Than The Previous Digit"**
~ **The Squaring of the Numbers Ending in 5** eg: 25^2
~ Optional or Advanced Study:
Sutra: **The First By The First and The Last By The Last**"
eg: 26x86

page 107

DAY 8: ~ **Sutra**: **Transpose & Apply** eg: 34x5
reviewed as 34x10 then ÷ by 2
~ Use other examples eg: 34x50 and 34x500 and 88x250
Compare this to Trachtenberg Method eg: 34x5

page 111

DAY 9: ~ **Sutra: "Vertically and Crosswise"**
~ **2 D**igits x **2 D**igits and **3 D**igits **x 3 D**igits eg: 123x345
~ **Developing The Inner Mental Screen**
(to be studied in next booklet known as Level 2:
- **3 D**igits **x 2 D**igits: The Moving Multiplier Method.
- **4 D**igits **x 4 D**igits)

~ **Gelosia**: How did The Arabs use Cross-Grids to solve Teen Numbers, all 2 D x 2 D Numbers and 3 D x 2 Digit Numbers.

page 128

DAY 10: ~ Revision Test: **Mathletics**, with Answers
~ End of Course

END Of RAPID MENTAL CALCULATION Workbook Level 1

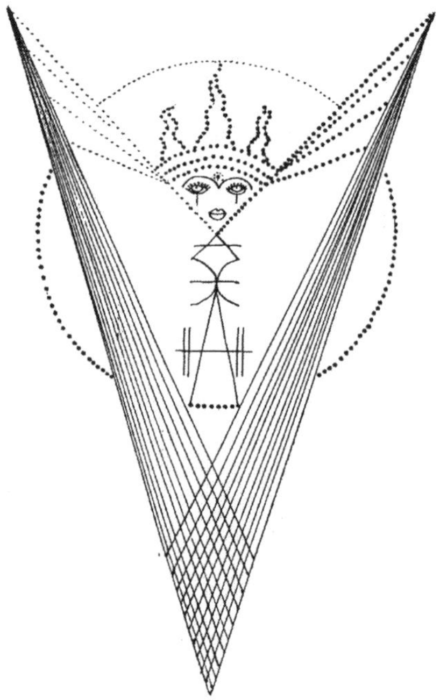

LEGEND:

aka = also known as
LHS = Left Hand Side
RHS = Right Hand Side
ibid = same as previous listing
2D x 2D = A 2 Digit Number Multiplied By Another 2 Digit Number
3D x 3D = something like 123 x 321
ie: = that is
eg: = for example

INTRODUCTION

NOTES At The BEGINNING Of COURSE:

Registration: Date ....

Venue: At ....

Name Introductions by throwing ball around: stating :

- My name is .... and my favourite fruit is ....
  This also teaches Memory Power by using an association with an Image.
- **Introduction** to Vedic Mathematics and Pattern Recognition
- **Revision** of the Days work.
- **Kumon** = Japanese concept of revising material within 24 hours to lock the information in your brain for life.

- **nb**: These lectures are leading to all day seminars on Sacred Geometry eg: ATOMIC STRUCTURE Of PLATINUM CRYSTAL BASED On The DIGITAL COMPRESSION Of The MULTIPLICATION TABLE.

Thus we introduce the Times Tables in this 10 week course, but during the school holidays, we go deeper into full day studies.

DAY 1

- **Magic Fingers**:
  - **9 Times Table Only,** looking at Back of Hands.
    Multiplication of a Single-Digit by Nine eg: 4x9
  - **9 Times a 2 Digit Number** eg: 9x38
  - The **Times Table**: **1 Digit x 1 Digit**
  - Plotting the **4 or 5 Times Table** in the **9 Point Circle** forming the **Enneagram**.
  - **Magic Fingers** Looking at Palm of Hands
    aka Universal Times Table solving 6x, 7x, 8x, 9x

- **Multiplication by 11** (Sutra: Digital Sums)
  - x11 any 2-Digit Number, eg: 34x11
  - x11 any 4-Digit Number, eg: 1,234x11 "Adding The Neighbour"
  - x11 Digit Sum Check for Divisibility by 11 (Optional)

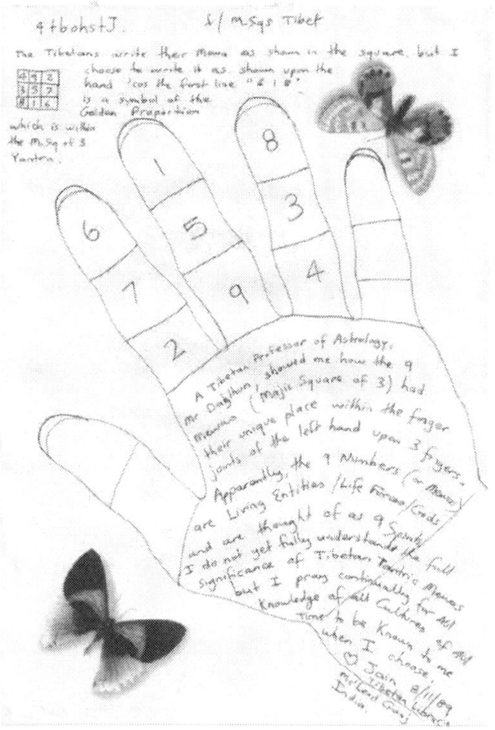

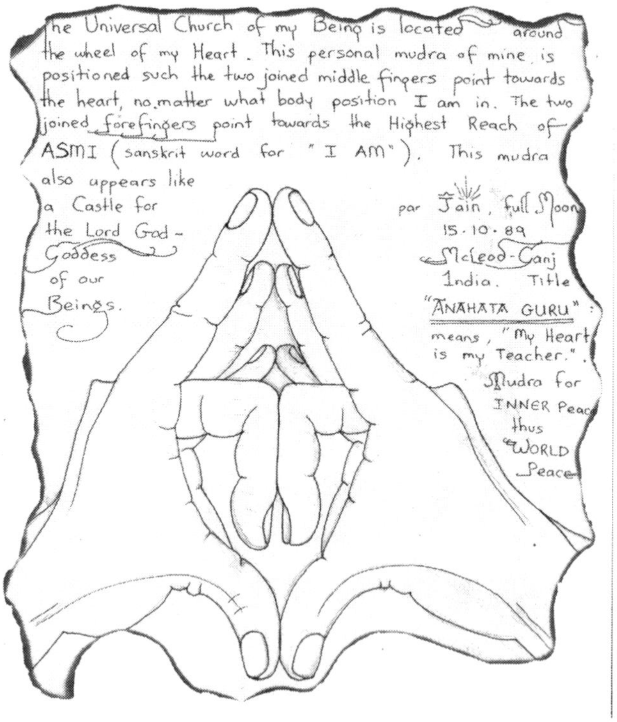

(drawings of Magic Hands by Jain 108 in 1989 whilst traveling in Tibet)

EXERCISE 1

Here is the 5 Times Table:

5 10 15 20 25 30 35 40 45 50 55 60 65 70

....

Upon the dotted line, add the above numbers so that they are reduced to a single digit

eg: 35 = 3 + 5 = 8

eg: 55 = 5 + 5 = 10 = 1 + 0 = 1

This is also known as the **Digital Compression or Digit Sums**
And is attributed to one of the 14 Sub-Sutras or formula.
As we discover these Sutras, we will systematically mark them where they are shown in the following Chapter 2.

What is the Periodicity of this Sequence? That is, how many numbers, as if on a necklace of beads, are repeating as the Digitally Compressed Sequence goes to Infinity.

....

Write out the Sequence of 9 Numbers that keep repeating, ready to plot them upon the 9 Point Circle.

....

This is a good introductory example of what I teach:
the ART OF NUMBER.

Now draw a long continuous Line of these 9 numbers, obeying the order of their progression. That is, draw a long unbroken line, from beginning to end, from 5 to 1 to 6 to 2 to 7 to 3 to 8 to 4 to 9 and then connect the end to the beginning.

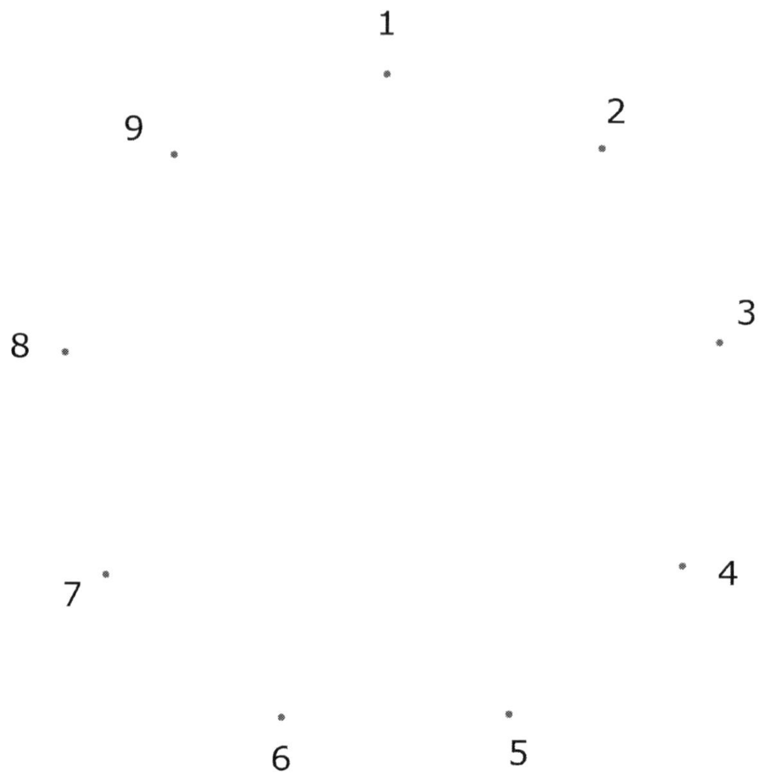

The Circle Of 9 Dots, Numbered from 1 to 9

In the space above,
draw the Digitally Compressed 5 x Table
upon the 9 Point Circle

ANSWERS 1

The Digital Compression of the 5 Times Table is:
5, 1, 6, 2, 7, 3, 8, 4, 9,   5, 1, 6, 2, 7, 3, 8, 4, 9,

It has a **Periodicity of 9**,
ie: 9 digitally compressed digits repeat infinitely.

The Pattern, or Yantra created when plotting these 9 digits upon the 9 Point Circle looks like this symmetrical and familiar pattern:

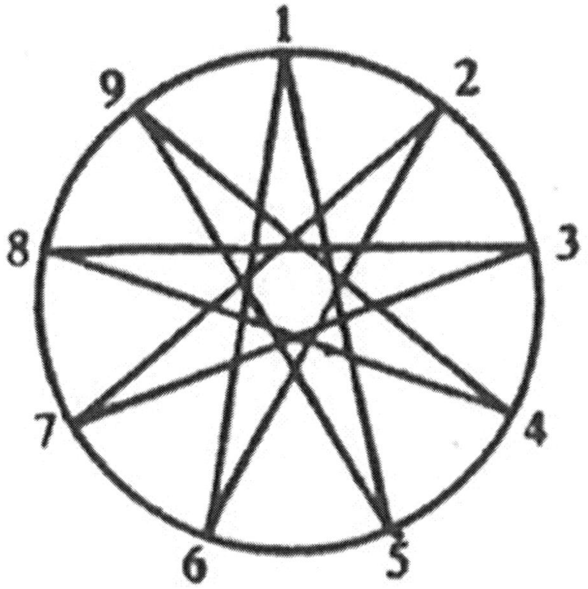

Technically it is called an Enneagram or 9 Pointed Star.
(When the 9 points are joined to form a 9-sided shape or form, it is called an Enneagon or Nonagon.

We can apply this Digital Compression to all of the Times Table, like the 2x or 3x or 4x etc and they all make exquisite patterns. It is known as **The Visual Multiplication Table Digitally Compressed**, and falsely known as The Vedic Square. It is not Vedic knowledge, it is Universal, and belongs to no culture. In other books that I have written, I have shown all the Times Table, from 1 to 9, commonly seen as a 9x9 grid, have compressed all the numbers to single digits,
eg: 8x8 = 64 and 64 = 10 = 1+ 0 = 1 then joined lines of connection to all similar numbers,
eg: what pattern would you get if you joined all the six digits of 1 that appear?
The dramatic and surprising solution is shown below:

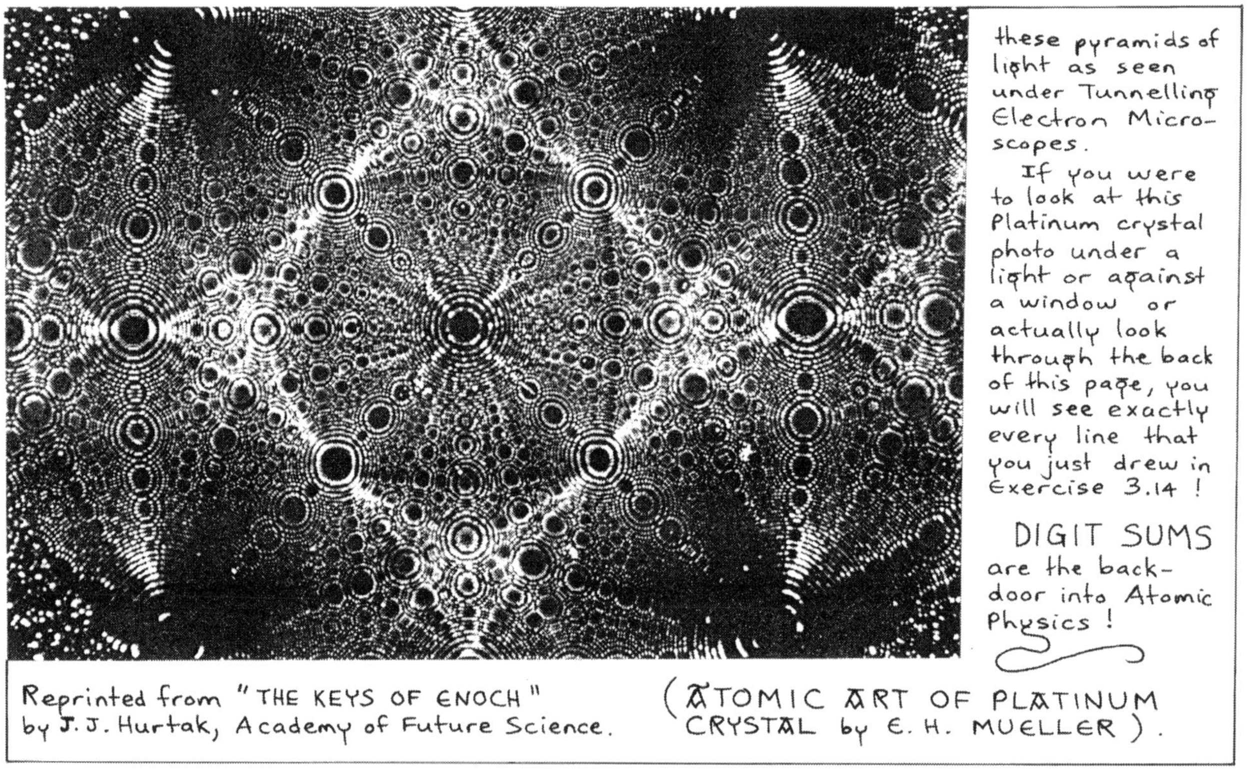

The **Atomic Structure of Platinum Crystal** that shows
"All The Digitally Compressed Points Of One" connected.
This will be drawn during a full day seminar.
(Original Crystal Atomic Photograph by E. H. Mueller.
Reprinted from "The Keys Of Enoch" by J. J. Hurtak).

9 TIMES TABLE USING THE BACK OF THE HANDS

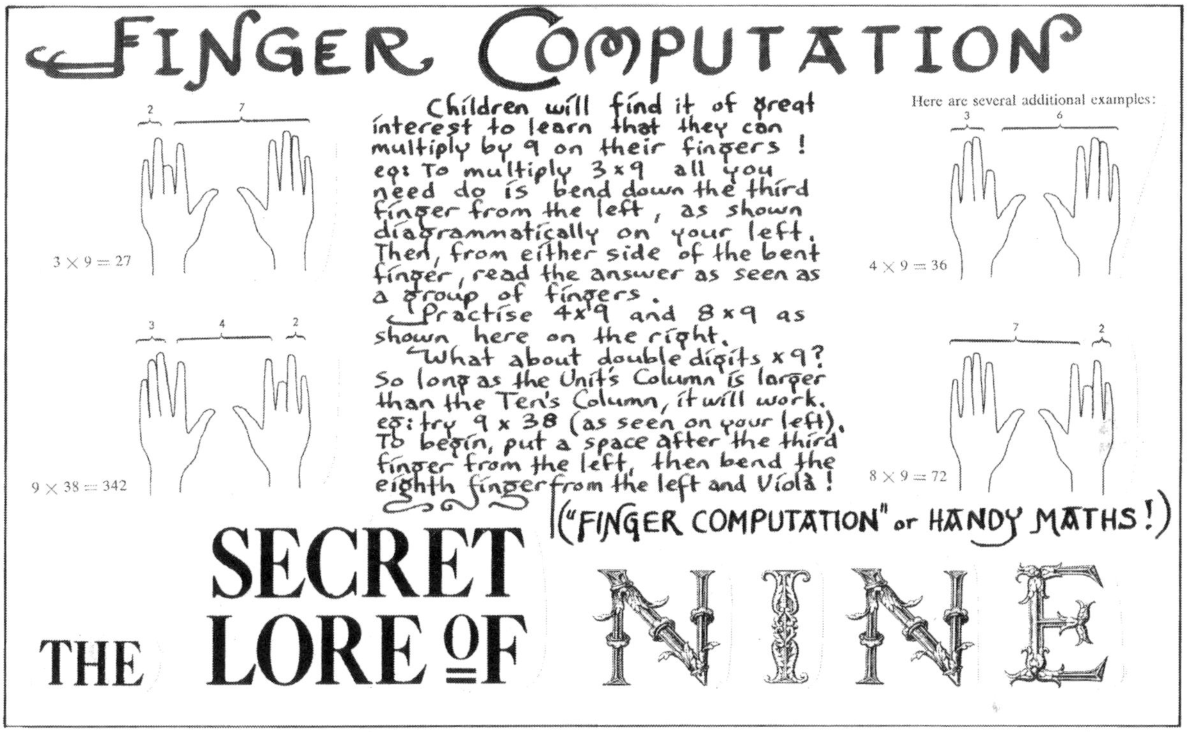

MULTIPLYING SINGLE DIGITS
BY NINE
USING FINGERS ONLY

(JAIN'S HANDY MATHS)

FINGER COMPUTATION

Children will find it of great interest to learn that they can compute their Times Table, like multiplying by 9, merely by glancing at their fingers!
Have your arms extended out so that you are viewing the back of your hands, with your ten fingers fanned out in front of you.

Focus now on your left hand, and starting from your left baby finger, give it a value of "1" and moving across to the left thumb, give each successive finger a value of increasing order. Other words for this are sequentially, or in consecutive order, or in natural counting order ie: ring finger = "2", middle finger = "3", index finger = "4", thumb = "5". Focus now on your right hand, and starting from your right thumb, give it a value of "6", and moving across to the left thumb, give each successive finger a value of increasing order, ie: index finger = "7", middle finger = "8", ring finger = "9", thumb = "10". It's as if you are looking at the fingernails and each nail has those numbers written on them in texta color.

To multiply 3 x 9, all you need to do is to count to your 3rd finger, starting from the far left and bend it down to almost touch the palm of that hand. Then from either side of the bent finger, read the answer which is revealed as two groups of fingers:
The two fingers on the left hand side represent the Tens Column, thus equals 20, and the 7 fingers on the right hand side represent the Units Column, thus equals 7. Combined, the instant answer is 20 + 7 = 27.

EXERCISE 2

Calculate, using your fingers, the multiplication of:

a) 7 x 9 =

b) 4 x 9 =

c) 8 x 9 =

d) 2 x 9 =

e) 6 x 9 =

f) 5 x 9 =

g) 9 x 9 =

MULTIPLYING DOUBLE DIGITS BY NINE, USING FINGERS ONLY, & LOOKING AT THE BACK OF YOUR HANDS

What ideas do you have regarding multiplying larger numbers by 9, like the double digit number 38?

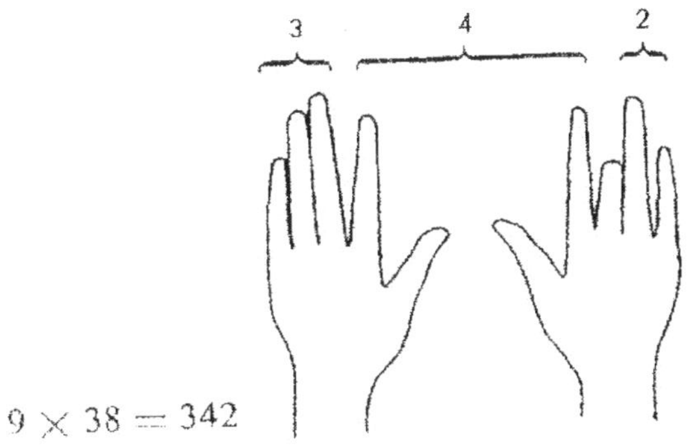

EXAMPLE 38 X 9

Double Digits
Multiplied by 9
using
Finger Computation
on Back Of The Hand:
Using the "V" and the "Bend"

To compute 38 x 9, have your arms and fingers extended out and locate the third finger from the left hand side. This is your left middle finger. This time, rather than bending that finger, make a space between this and the fourth finger, like a "**V-shape**".

Regarding the "8" of the "38", focus now on the eighth finger which is the middle finger of your right hand and **bend** this one down. Voila! You now have your instant answer before your eyes.
You will observe that there are 3 compartments or groupings of your fingers formed by the "**V-shape**" and the **bent finger**.
Beginning from the left hand side, call out these groupings:
3 / 4 / 2, giving an answer of 342.

EXERCISE 3

Calculate, using your fingers, the multiplication of:

a) 9 x 47 **b)** 9 x 39 **c)** 9 x 26
d) 9 x 24 **e)** 9 x 17 **f)** 9 x 56

LIMITATION OF THIS METHOD:
nb: There is an anomaly or fault in this system.
We have been taught that the Multiplication of Double Digits by 9, using the special finger technique, can only work out correctly when the Units Column is larger than the Tens Column
eg: 9 x **3**7 can be solved, as **3** is less than 7,
but not 9 x 73 as 7 is greater than the 3.
The mathematical symbols for "lesser than" is " **<** "
and the symbol for "greater than" is " **>** ".
To rewrite the above: eg: 9 x 37 can be solved, as 3 < 7,
but not 9 x 73 as 7 > 3.
The new method, explained in my book: "Jain Mathemagics Curriculum For The Global School, Book 2, Multiplication". Please Read: Chapter 163 "Postscript" on pages 148.

ANSWERS 3

a) 423 **b)** 351 c) 234 **d)** 216 **e)** 153 **f)** 504

Philosophy is written in this grand book - I mean the universe - which stands continually open to our gaze, but it cannot be understood unless one first learns to comprehend the language and interpret the characters in which it is written.
It is written in the language of mathematics, and its characters are triangles, circles, and other geometrical figures, without which it is humanly impossible to understand a single word of it; without these, one is wandering about in a dark labyrinth.

-Galileo Galilei
Il Saggiatore [1623]

MAGIC FINGERS

or "HANDY MATHS"

UNIVERSAL TIMES TABLE SOLVING 6X, 7X, 8X 9X

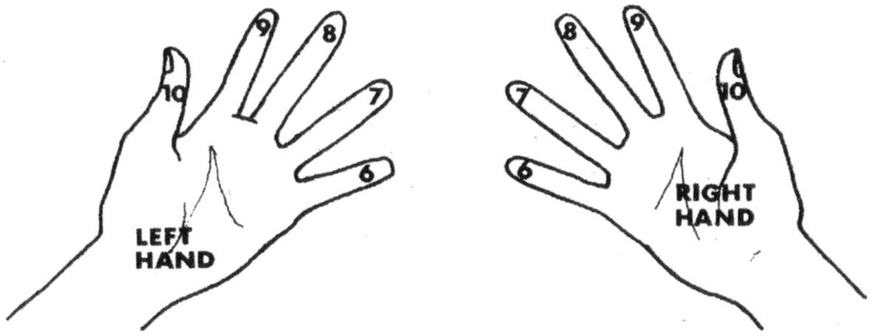

The UNIVERSAL TIMES TABLE

Achieved by using the fingers on each hand that represent the numbers from 6 to 10

There is a special technique for knowing your Times Table, quickly, by mere use of finger dexterity. It use to appear in old mathematical text books and unfortunately became obsolete, but here it is revived to do it the justice it deserves.
I went around the world in 2004 and 2005 teaching "Magic Fingers" and was astounded by how popular it had become. Every teacher to whom I have taught it to, loves it. It is quite a thrill to learn Magic Fingers in 15 minutes, and realize that for another child to learn their 6x, 7x, 8x and 9x Tables, it would traditionally take them 2 years during their primary schooling!

Referring to the figure above, mark the fingers of both hands, in texta color, or imagine they are written upon them, with the numbers 6, 7, 8, 9, and 10 beginning with the baby finger as No. 6, the ring finger as No. 7 etc till you get to the thumb as No. 10.

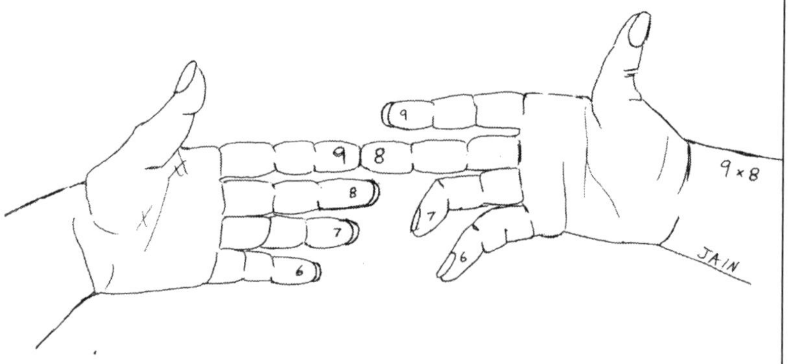

Multiplying
9 X 8
Using Fingers Only, Facing The Palms Of Your Hands.

LEARN 2 YEARS OF ROTE LEARNING LABORIOUS TIMES TABLES IN 15 MINUTES! AMAZING.

Suppose you require to know what **9 x 8** is?

Observe the diagram above.
Locate on your Left Hand the finger representing 9, which is the index finger, and locate on your Right Hand the finger representing 8, which is the middle finger.
With palms facing you, join those 2 fingertips to touch and allocate to each of the joined fingers, plus the 5 fingers below them, a value of 10 units. All together there are 7 lots of 10.

This gives a value of 70. Add to this, the multiplication of the remaining fingers. Observe that there is 1 finger on the Left Hand that is above the joined fingers, and there are 2 fingers surplus or remaining on the Right Hand. Multiplying these 2 excesses or remainders is 1 x 2 = 2. The final answer is therefore 70 + 2 = 72.
Children adore this method, as they no longer need to rote learn their 6 Times, 7 Times, 8 Times and 9 Times Tables, the ones that they often have difficulty with. It gives them a connection with numbers that is tactile, tangible, simple. There is a sense of achievement and wonder from the nimble and dexterous activity happening in front of their eyes.
Highly recommended to learn, so practice with these following examples:

EXERCISE 4

Use your fingers to multiply the following Single Digits X Single Digits:

a) 8 x 9 =
b) 7 x7 =
c) 7 x 9 =
d) 8 x 8 =
e) 6 x 7 =
f) 6 x 9 =
g) 6 x 8 =
h) 6 x 6 =
i) 9 x 9 =
j) 8 x 7 =

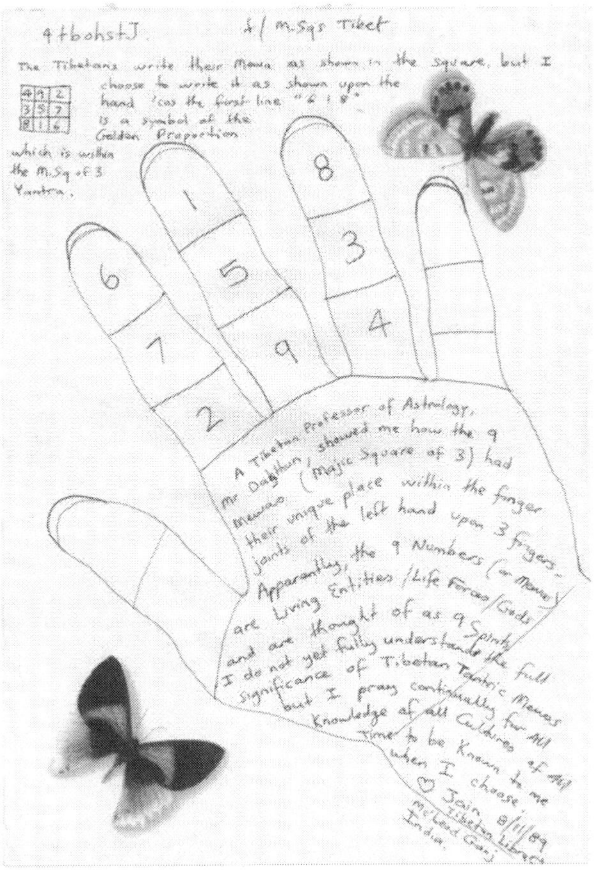

ANSWERS 4

a) 72 **b)** 49 **c)** 63 **d)** 64 **e)** 42
f) 54 **g)** 48 **h)** 36 **i)** 81 **j)** 56

DIGIT SUMS
MULTIPLICATION BY 11

SPLITTING THE DIGITS AND ADDING THEIR SUM

Here is a clever short-cut that involves Digit Sums and develops number skills.
To multiply any 2-digit no. by 11, just split the digits apart and insert between them the sum of those 2 digits :

Eg : multiply **33 x 11**

= 3 (3+3) 3
= 3 6 3

Eg : Multiply **71 x 11**

= 7 (7+1) 1
= 781

EXERCISE 5

Mentally Multiply by 11 the following numbers :

a) 21 =

b) 35 =

c) 42 =

d) 52 =

e) 63 =

f) 71 =

g) 81 =

h) 90 =

Notice that the Digit Sums were all "< 9" meaning "less than 9". When the Digit Sum is "> 9" (Greater Than Nine) you will have a 2-Digit teen number. Leave the unit's digit between the 2 split digits and carry the 1 to the left to the first digit of the answer.

EXAMPLE 48 x 11

=	4	(4 + 8)	8
=	4	12	8
=	4	$_1 2$	8
=	5	2	8

(a shortcut for writing this is:

48x11= 4/12/8 = 5/2/8 = 528)

EXERCISE 5a

Mentally Multiply by 11 the following numbers :

a) 38 =

b) 28 =

c) 49 =

d) 56 =

e) 66 =

f) 79 =

g) 87 =

h) 99 =

ANSWERS 5

a) 231
b) 385
c) 462
d) 572
e) 693
f) 781
g) 891
h) 990

ANSWERS 5a

a) 3/11/8 = 418
b) 2/10/8 = 308
c) 4/13/9 = 539
d) 5/11/6 = 616
e) 6/12/6 = 726
f) 7/16/9 = 869
g) 8/15/7= 957
h) 9/18/9 = 1089

HOW To MULTIPLY By 11 NUMBERS LARGER THAN 2 DIGITS
+
MULTIPLICATION of a 4 DIGIT NUMBER By ELEVEN

Sutra: ADD the NEIGHBOUR on the RIGHT
This is the secret to multiplying larger nos. by 11

EXAMPLE 1,234 x 11

1) Add a Zero in front and at the end of the number :
   = **0** 1 2 3 4 **0**

2) Starting from the Right Hand Side, and progressing to the left. Drop the last Digit (which is 4) or add the Neighbour (which is 0) to the right of the last digit
   = 4 + 0 = **4**
   written as 0123**4**0
   4 **nb:** the "4"s exactly lined up.

3) Now go to the next digit on the left which is **3**.
   Add the Neighbour on the right
   = 3 + 4 = **7**
   written as 012**3**40
   74

4) Now go to the next digit on the left which is **2**. Add the Neighbour
   = 2 + 3 = **5**
   written as 01**2**340
   574

5) Now go to the next digit on the left which is **1**. Add the Neighbour
   = 1 + 2 = **3**
   written as 0**1**2340
   3574

6) Now go to the next digit on the left which is **0**. Add the Neighbour
   = 0 + 1 = **1**
   written as **0**12340
   13574

Thus 1234 x 11 = **13,574**

Here are 2 more examples with carry-overs.

Beginning from the right hand side of the number being multiplied, I have hand written them to show how I systematically add the neighbour to the right, using little semi-circles above the numbers, and with the carry-overs, they are placed carefully to the left, upon another lower line.

EXAMPLE 111,358 x 11

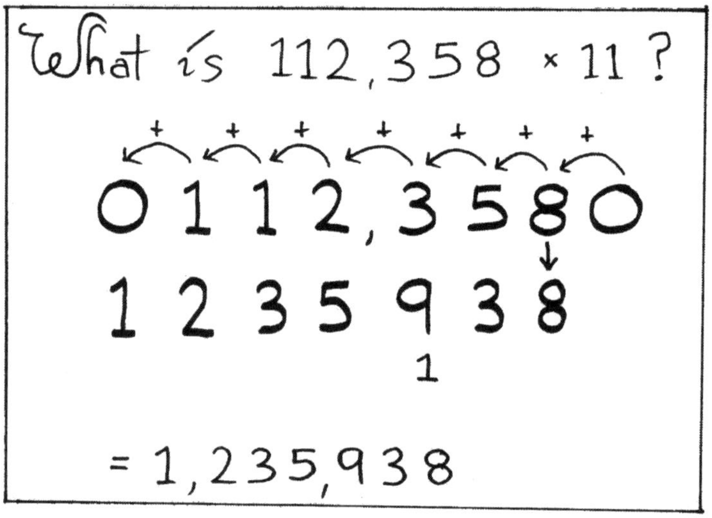

EXAMPLE 24,593 x 11

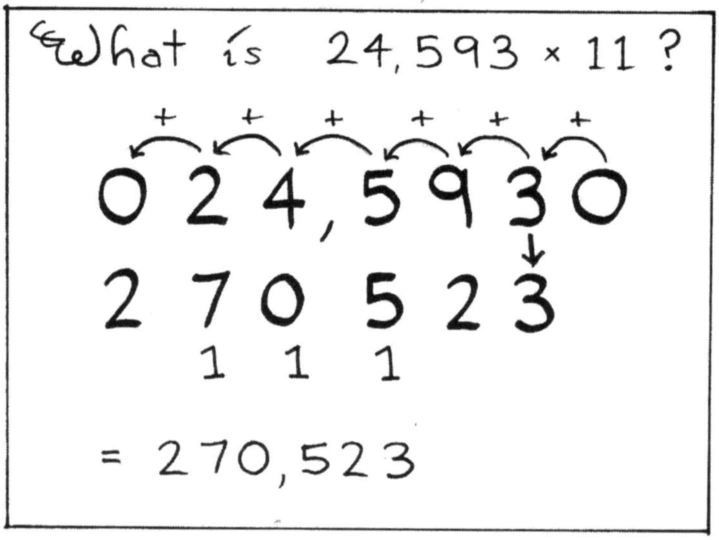

EXERCISE 6

MULTIPLY BY 11

a) 123 =

b) 2,453 =

c) 4,162 =

d) 58,024 =

e) 112,358 =

f) 24,593 =

ANSWERS 6

a) 1,353 b) 26,983 c) 45,782 d) 638,264
e) 1,235,938 f) 270,523

[Notice in this last example, when 'Adding the Neighbour' of 5+8 gave a Digit Sum in the teens, which meant there was a carry–over to the next Digit on the left].

(This Worksheet is Optional)

DIGIT SUM CHECK for MULTIPLICATION by ELEVEN or TESTING for DIVISIBILITY by 11

Let us check that **135,795** is divisible by 11

It is good practice to have all your calculations checked by Digit Sums. Presented here is another oddity.

1) Underline every 2nd No. beginning from the last digit on the right
   1 3 **5** 7 **9** 5

2) Add the digits underlined: 1 + 5 + 9 = 15

3) Add the digits not underlined: 3 + 7 + 5 = 15

4) The answer is correct when both digit sums are the same (i.e. 15 = 15) !
   Sometimes they differ by a multiple of 11 (this means the differences between the two Digit Sums can be 11, 22, 44, 55 etc).

EXERCISE 6a

- Check that the Digit Sums, for the following numbers, prove DIVISIBILITY by 11 by answering YES or NO.

a. 356,576 =

b. 38,731,781 =

c. 252,646 =

d. 986,194 =

ANSWERS 6a

a. **3**5**6**,5**7**6 sums are 3+6+7=16 and 5+5+6=16 therefore: YES

b. **3**8,**7**3**1**,7**8**1 sums are 3+7+1+8=19 and 8+3+7+1=19. YES

c. **2**5**2**,6**4**6 sums are 2+2+4=8 and 5+5+6=17 and the difference between the two sums is 17-8=9 which is not a multiple of 11 or zero, therefore: N0

d. **9**8**6**,1**9**4 sums are 9+6+9=24 and 8+1+4=13 and the difference between the two sums =24-13=11, therefore: YES

(Graphic from Solara, front cover of her book: "11:11", when she visited Byron Bay on a world tour promoting the next major planetary activation on January 11, 1992)

DAY 2

- **Chart** of the **16 Sutras** and **14 Sub-Sutras**.
- About **Bharati Krsna Tirthaji**
- **Complements of Nine** (eg: 1+8, 2+7, 3+6)
- **Complements of Ten** (eg: 1+9, 2+8, 3+7)
- **Sutra: "All From Nine And The Last From Ten"**

MAKING A NOTE OF THE SUTRA OR SUB-SUTRA EMPLOYED

After the completion of each day lesson, make a summary of what were the major sutras or formulae taught and make a note of this by referring to the chart that lists both
1 – **16** Vedic Mathematics **Sutras**
2 – **14** Vedic Mathematics **Sub-Sutras**.

According to the previous DAY 1 Lesson, the main sutra covered was called DIGIT SUMS or DIGITAL COMPRESSION but was known 5 decades ago as
THE SUM OF THE PRODUCTS, and its Sanskrit name is: SAMUCCAYA-GUNITA.
This sub-sutra was used in the Multiplication of 11, as in 25x11 = 2/**2+5**/5 = 275 where digits were added to achieve an instant answer.
The teacher must now guide the students to mark this special sub-sutra by referring to the page that says: "Sub-Sutras" and ticking the number "10" that refers to "The Sum of the Products".

You can progressively mark your Sutras with a tick or an arrow or any other creative squiggle that shows you that you have read and understood the formula.

CHART OF THE 16 MAIN AND 16 SUB-SUTRAS

16 SUTRAS	14 SUB-SUTRAS
1 – By One More Than the One Before	1 - Proportionately
2 – All from 9 and the Last from 10	**2 – The Remainder Remains Constant**
3 – Vertically and Cross-wise	**3 – The First by the First and the Last by the Last**
4 – Transpose and Apply	4 - For 7 the Multiplicand is 143
5 - If the Samuccaya is the Same it is Zero	5 - By Osculation
6 - If One is in Ratio the Other is Zero	**6 – Lessen by the Deficiency**
7 - By Addition and by Subtraction	**7 – Whatever the Deficiency lessen by that amount and set up the Square of the Deficiency**
8 – By the Completion or Non-Completion	8 - Last Totalling 10
9 - Differential Calculus	9 - Only the Last Terms
10 – By the Deficiency – By The Excess	**10 – The Sum of the Products** aka **Digit Sums** or **Digital Compression**
11 - Specific and General	11 - By Alternative Elimination and Retention
12 - The Remainders by the Last Digit	**12 – By Mere Observation**
13 - The Ultimate and Twice the Penultimate	13 - The Product of the Sum is the Sum of the Products
14 - By One Less than the One Before	**14 – On the Flag**
15 - The Product of the Sum	**15 – Trachtenberg** 2Digit x 1 Digit
16 - All the Multipliers	**16 – Add The Neighbour**

nb: I have added the 15th Sub-Sutra: "Trachtenberg" to honour this method.
nb: I have also added the 16th Sub-Sutra: "Add The Neighbour" that solves Multiplication by 11 with 3 Digit Numbers, 4 Digit Numbers and more.

Notice how the relevant **10th Sub-Sutra: "DIGIT SUMS"** has been highlighted.
Of these 30 formulae, 15 have been marked in bold and they are the actual ones that will be dealt with in this 10 week course according to the teachers time management in the School Term.
If this Level One Course is called Beginners, then subsequent courses like Level 2 (Intermediary) and Level 3 (Advanced) will deal with most of the others sutras. Do not worry about these complicated names, they will all appear to make sense by the end of the course.

And just for your reference, here is the same chart of the 16 main Sutras showing how they were written in the original **Sanskrit** or Indian language:

VEDIC MATHEMATICS SUTRAS

1	एकाधिकेन पूर्वेन Ekādhikena Pūrvena	*By One More than the One Before*
2	निखिलं नवतश्चरमं दशतः Nikhilam Navataścaramam Daśatah	*All from 9 and the Last from 10*
3	ऊर्ध्वतिर्यग्भ्यामं Ūrdhva Tiryagbhyām	*Vertically and Cross-wise*
4	परावर्त्य योजयेत् Parāvartya Yojayet	*Transpose and Apply*
5	शून्यं साम्यसमुच्चये Sūnyam Sāmyasamuccaye	*If the Samuccaya is the Same it is Zero*
6	आनुरूप्ये शून्यं अन्यत् (Ānurūpye) Sūnyamanyat	*If One is in Ratio the Other is Zero*
7	संकलन व्यवकलनाभ्यां Saṅkalana Vyavakalanābhyām	*By Addition and by Subtraction*
8	पूरणापूरणाभ्यां Pūranāpūranābhyām	*By the Completion or Non-Completion*
9	चलनकलनाभ्याम् Calana Kalanābhyām	*Differential Calculus*
10	यावदूनं Yāvadūnam	*By the Deficiency*
11	व्यष्टिसमष्टिः Vyastisamastih	*Specific and General*
12	शेषाण्यङ्केन चरमेण Sesānyaṅkena Caramena	*The Remainders by the Last Digit*
13	सोपान्त्यद्वयमन्त्यं Sopāntyadvayamantyam	*The Ultimate and Twice the Penultimate*
14	एकन्यूनेन पूर्वन Ekanyunena Purvena	*By One Less than the One Before*
15	गुणितसमुच्चयः Gunitasamuccayah	*The Product of the Sum*
16	गुणकसमुच्चयः Gunakasamuccayah	*All the Multipliers*

The Author
Jagadguru Śaṅkarācārya
Śrī Bhāratī Kṛṣṇa Tīrthajī Mahārāja
(1884—1960)

Bharati Krsna Tirthaji (1884 - 1960)

(The Ecclesiatical Head or Pope of a quadrant of India, also an esteemed Mathematician)

SRI SHANKARACHARYA AT SRF-YSS HEADQUARTERS

Sri Jagadguru Shankaracharya Bharati Krishna Tirtha of Puri, India, at Los Angeles international headquarters (founded in 1925 by Paramahansa Yogananda). In 1958 the senior head of the Swami Order paid a three-months' visit to America, sponsored by Self-Realization Fellowship. It was the first time in the history of the ancient Swami Order that a Shankaracharya had traveled to the West.

(extract from "Autobiography of a Yogi" by Yogananda Paramahansa)

Paramahansa Yogananda introduced yoga to the west. He formed the Self Realization Fellowship, and died around 1952. Six years later, the head of this organization, a female devotee, organized Bharati Krsna Tirthaji's tour.

Biographical History Surrrounding Bharati Krsna Tirthaji's Life

- **1884** - born to highly learned and pious parents and known as Venkatraman.
- Was dux of every school he attended in teenage years eg at Madras University and excelled with highest honours in 8 subjects: Sanskrit, Philosophy, English, Science, History, Mathematics.
- His mentor was Bhartrhari – a philosopher of grammar.
- Gave many talks of Vedic Maths in many universities in India (and later in USA).
- **1908** – A burning desire for spiritual knowledge, begins a monastic life but was swayed to do 3 years as a principal of a college.
- **1911** – returns back to a monastic life for 6 years in deep study of Vedanta, Brahma-Sadhana and vigorous practices in a nearby forest in Sringeri, leaving only sometimes to teach at Poona, on request.
- **1919** - Initiated in the highest order of Samnyasa at Varanasi (Benares) by his guru, and changed his name to Bharati Krisna Tirtha, age 35.
- As an acclaimed authority on Sthapathyaveda (the mathematics of Engineering and Temple Building) he tours India and took the entire intellectual and religious class of the nation by storm, spreading Sanatana Dharma which promotes Vedic concepts of world peace. His guru has failing health and insists the Swami be Head of this Organization.
- **1956** – Confirmed loss of 16 volumes of Vedic Mathematics, that he wrote on the 16 Sutras, and were stored at a disciple's house. Some Indians copying this Rapid Mental Calculation had already dazzled audiences, claiming occult powers without acknowledgement to the master's teachings, therefore his closest friends insist him to arrange publication of the 16 sutras in his own name (but the one book, a mere introduction of 400 pages, that he writes in 1957 gets published 7 years posthumously).
- **1957** – He easily writes via dictation the only one surviving book, having an astute memory but failing eyesight. (This book is to become the bible on Mathematics in the western world 50 years posthumously).

continued...

➢ **1958** – Tours America. This is a historical moment, as it is the first time that a Shankaracharya of this High Order left India (known really as "Bharat") to go abroad.
(Usually there are brahmanic rules that state that a holy man is never allowed "to cross the waters" etc, but he believed that to raise planetary consciousness it was important to defy the scriptures and release the 16 sutras or threads of knowledge that had been lost. Unfortunately this knowledge gets buried by the American intellectual Gestapo who do not give these mathematical gems to their culture, but bury it for many decades until people like Mr Kenneth Williams of England and Jain of Australia empower their students with the revival of this material. NASA use it secretly in the field of Artificial Intelligence, in the sense of how does a robot's brain works with mathematical calculations).

➢ **1958** – was sponsored by the Self Realization Fellowship in Los Angeles, the Vedantic Society founded by Paramahansa Yoganandaji who died in 1952, and succeeded by Mata. Bharati Krsna spends 3 months lecturing in universities, radios, and also made some television appearances. His profound knowledge of Mental, One-Line Arithmetic takes the intellectual world by storm.
(American officials who witnessed these lectures realizes that if every child learnt this system, they would have a whole race of geniuses, and this can not be allowed in any plutocracy. As the years go by, it is obvious that there is no available record of all the film transcripts of this most important visit, and it is suspected that the Vatican stole this archival footage, as well as the original 16 hand written volumes of books that went missing years prior to this tour).
(Indian publishers offer a huge reward for the retrieval of these books and film footage in the year 2004 and set up a society to reclaim their lost heritage, and to relinquish the centuries of British control of the Indian Mathematical Curriculum).

➢ **1960** – Over-worked and ill-health sets in; Leaves his body.

COMPLEMENTS OF 9 AND 10

One of the essential requirements to Rapid Mental Calculation is to know your Complements of 9 and 10.
I suggest to playful ways to entrench the memory of these Pairs:

1 – Auditory style of teaching: Sing songs to your child allowing them to give a response that tests their knowledge of these Pairs.

eg: "10 eggs in a nest
snakes ate 5,
How many left?"

eg: "9 chokos on a vine
the chooks ate 5,
How many left?"

Really the student needs to know the Pairs not just of 9 and 10, but all numbers below these, as in:

eg: "7 cassowarys on the track
3 ran away,
How many left?"

eg: "5 water melons in the garden
2 were sold,
How many left?"

2 – Visual style of teaching:
To Teach all the Pairs of 9:
Select from your standard deck of playing cards all the numbers from 1 to 9. Also include the 4 Jacks or any of the 4 picture cards to act as the zeroes. Delete the remaining 8 picture cards, delete the 4 tens and the Joker card if it is there.
Place all 40 cards face down, after being shuffled. Two or more students can play, having turns to turn over 2 cards, aiming to gets Pairs like
0 and 9,
1 and 8,
2 and 7,
3 and 6,
4 and 5.

When a student turns over a correct pair, they can have another turn.
This game or exercise is more about Memory Power rather than Competitiveness.
You may then want to teach them the Pairs of 10.

To Teach all the Pairs of 10:
Select from your standard deck of playing cards all the numbers from 1 to 10. Also include the 4 Jacks or any of the 4 picture cards to act as the zeroes. Delete the remaining 4 Queens and 4 Kings, and delete the Joker card if it is there.
Place all 44 cards face down, after being shuffled. Two or more students can play, having turns to turn over 2 cards, aiming to gets Pairs like
0 and 10,
1 and 9,
2 and 8,
3 and 7,
4 and 6,
5 and 5.
When a student turns over a correct pair, they can have another turn.

MAGIC FINGERS:

TEACHING CHILDREN THE COMPLEMENTARY PAIRS OF 9

The diagram below is a beautiful representation of how the 10 Fingers actually represent the underlying code of 9,
and accentuating the Hindu invention of Zero.
Young children like it when you write these numbers on their fingers with permanent markers, and they can see for themselves, the pairs that add up to 9.

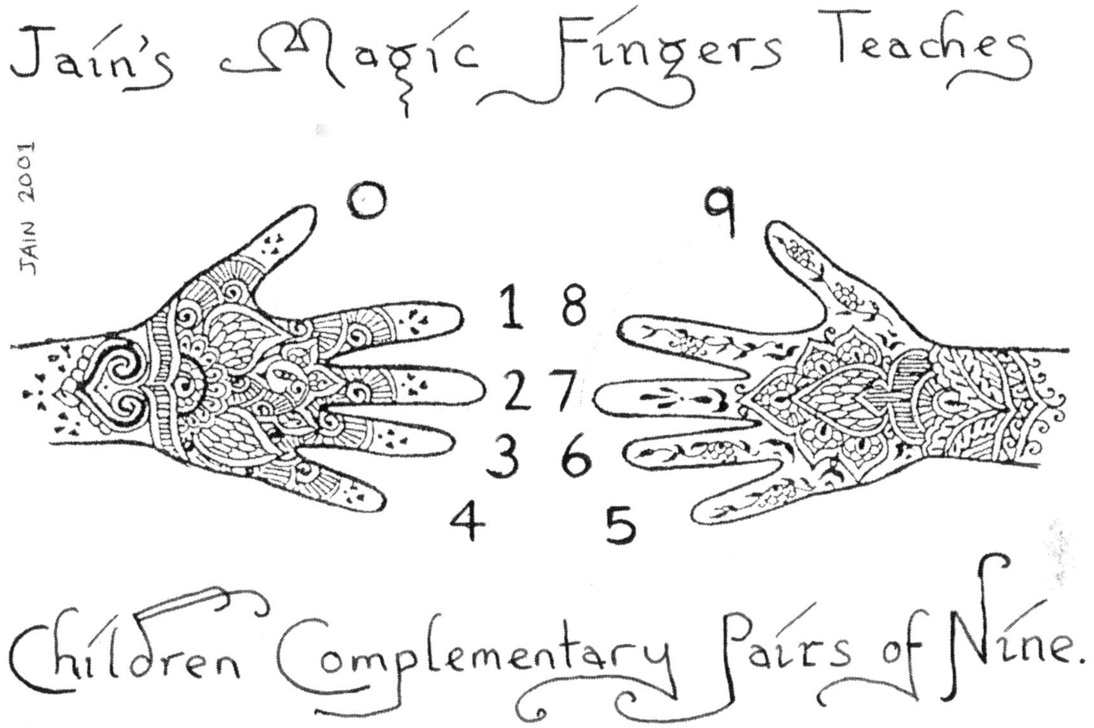

HOW DO I BEGIN TO TEACH MY CHILD THE BASIC FIRST PRINCIPLES OF VEDIC MATHEMATICS?

Start by singing songs to your child that will educate them with a fun understanding of all the Pairs of 9 (and all the Pairs of 10). eg: 9 Kookaburras sitting on a branch, 3 flew away, how many left?

(You will be surprise how many teenagers in the school system do not know their Pairs of 9).
That is why this program that teaches from the very beginnings the art of knowing your Pairs of 9 and Pairs of 10 and also the Times Table, are the two main pillars or foundations of all Mathematics. I highly emphasize that all students begin with this in mind.

The Sutra: **All From 9 And The Last From 10** is based on this Knowledge, and years down the track, these Pairs of 9 and 10 will help them solve larger mathematical calculations, mentally like:

998 x 678

(means I lessen 678 by the deficiency of 998 from 1,000 which is 002,
giving half of the answer as 676 / _ _ _
(Employing the Sutra: By The Deficiency)
and the other half of the answer is achieved
by knowing the Sutra:
All From 9 And The Last From 10
to determine the remainder of 678 from a 1,000
which is 9-6 / 9-7 / 10-8 = 322
which must be doubled because our deficiency was 002
giving 322 x 2 = 644
Thus the instant answer is: 676 / 644
which is 676,644).
Magic!

SUTRA
"ALL FROM NINE AND THE LAST FROM TEN"

In ancient Sanskrit, this process of Multiplication had a special name, it is called:

NIKHILAM NAVATASCARAMAM DASATAH

Which translates as: **ALL FROM NINE AND THE LAST FROM TEN**.

In India, it is shortened to NIKHILAM, meaning general Multiplication.

To truly learn Vedic Mathematics it is important that the student understands the **COMPLEMENTS OF NINE AND TEN**.

That means, regarding 9, that the student know all the Pairs that sum up to Nine, like:

1 + 8 = 9, 2 + 7 = 9, 3 + 6 = 9, 4 + 5 = 9 etc

and regarding 10,

that the student know all the Pairs that sum up to Ten, like:

1 + 9 = 10, 2 + 8 = 10, 3 + 7 = 10, 4 + 6 = 10, 5 + 5 = 10 etc.

In Steiner Schools (aka Waldorf Schools in Europe) the Pairs of 9 are referred to as **THE FRIENDS OF NINE**.

We will need to know these Pairs when we begin to multiply SINGLE DIGITS BY SINGLE DIGITS like 7 x 8 and 9 x 6 etc which is really an understanding of your MULTIPLICATION TABLE.

Before we do this, let us practice some examples of ALL FROM NINE AND THE LAST FROM TEN, looking at quick mental subtractions from Base Numbers like 100, and 1,000 and higher.

EXAMPLE 100 minus 68

To subtract a number from 100, like **68** we say mentally, like a ritual "All From Nine And The Last From Ten means"
9 minus **6** / 10 minus **8** = 3 / 2 = 32 If we had to set it out on paper, it would look like this:

9	10
minus	minus
6	**8**
equals	equals
3	2

To write this out in the one-line answer format required it would look like this:
100 – 68 = 9 – 6 / 10 – 8 = 3/2 = 32

EXAMPLE 1,000 minus 784

To subtract a number from 1,000, like 784 we say:
All From Nine and the Last from 10 as a One-Line Answer:
9 minus 7 / 9 minus 8 / 10 minus 4 = 2 / 1 / 6 = 216
If we had to set it out on paper, it would look like this:

9	9	10
7	**8**	**4**
2	1	6

By mentally committing to memory all your complements of 9 and 10, you will be able to, within seconds, or by mere observation, subtract any number from say a million, effortlessly.

EXAMPLE Subtract from 1,000,000 the number 851,651

= 9 minus 8 / 9 minus 5 / 9 minus 1 / 9 minus 6 / 9 minus 5 / 10 minus 1
= 1 / 4 / 8 / 3 / 4 / 9
= 148,349

nb: if a number ends in zero, like 637,120 and this is subtracted from 1,000,000 to derive the complement, you say "All From Nine and the Last from Ten" but disregard the 0 of 637,120 and subtract the last digit 2 from 10.

EXERCISE 1

Subtract the following numbers from their nearest round base:

From 100

a. **36**

b. **52**

c. **88**

from 1,000

d. **280**

e. **488**

f. **716**

from 10,000

g. **0,199**

h. **5,904**

i. **8,366**

from 100,000

j. **29,319**

k. **65,575**

l. **82,704**

from 1,000,000

m. **385,344**

n. **637,120**

o. **838,949**

(design by Jain 108, hand-drawn:
a Magic Square of 7x7 rotated
upon itself 8 times)

ANSWERS 1

a.	100 - 36 = 9 - 3 / 10 - 6	= 6/4 = **64**		
b.	100 - 52 = 9 - 5 / 10 - 2	= 4/8 = **48**		
c.	100 - 88 = 9 - 8 / 10 - 8	= 1/2 = **12**		
d.	1,000 - 280	= 9 - 2 / 10 - 8 / 0	= 7 / 2 / 0	= **720**
e.	1,000 - 488	= 9 - 4 / 9 - 8 / 10 - 8	= 5 / 1 / 2	= **512**
f.	1,000 - 716	= 9 - 7 / 9 - 1 / 10 - 6	= 2 / 8 / 4	= **284**
g.	10,000 - 0,199	= 9 - 0 / 9 - 1 / 9 - 9 / 10 - 9	= 9 / 8 / 0 / 1	= **9,801**
h.	10,000 - 5,904	= 9 - 5 / 9 - 9 / 9 - 0 / 10 - 4	= 4 / 0 / 9 / 6	= **4,096**
i.	10,000 - 8,366	= 9 - 8/ 9 - 3 / 9 - 6 / 10 - 6	= 1 / 6 / 3 / 4	= **1,634**
j.	100,000 - 29,319	= 9 - 2 / 9 - 9 / 9 - 3 / 9 - 1 / 10 - 9	= 7 / 0 / 6 / 8 / 1	= **70,681**
k.	100,000 - 65,575	= 9 - 6 / 9 - 5 / 9 - 5 / 9 - 7 / 10 - 5	= 3 / 4 / 4 / 2 / 5	= **34,425**
l.	100,000 - 82,704	= 9 - 8 / 9 - 2 / 9 - 7 / 9 - 0 / 10 - 4	= 1 / 7 / 2 / 9 / 6	= **17,296**
m.	1,000,000 - 385,344	= 9 - 3 / 9 - 8 / 9 - 5 / 9 - 3 / 9 - 4 / 10 - 4		
		= 6 / 1 / 4 / 6 / 5 / 6 = **614,656**		
n.	1,000,000 - 637,120	= 9 - 6 / 9 - 3 / 9 - 7 / 9 - 1 / 10 - 2 / 0		
		= 3 / 6 / 2 / 8 / 8 / 0 = **362,880**		
o.	1,000,000 - 838,949	= 9 - 8 / 9 - 3 / 9 - 8 / 9 - 9 / 9 - 4 / 10 - 9		
		= 1 / 6 / 1 / 0 / 5 / 1 = **161,051**		

DAY 3

- **Angles** of the **9 Sanskrit Numbers**
  "**Al-Arquan–Al–Hindu**" means "The Hindu Figures" regarding the **Origin of where the numbers came from.**
  Show how the number **"3" originally had "3" angles etc.**

- **Vedic Addition**: Sutra: "**By The Completion**"

- How the **Indians** did their **Times Table** 2000 years ago using the Sutra: "By The Deficiency" for "Under Base 10" eg: 8x9

- **Addition Trick (4,090)** by adding:
  1,000 + 10 +1,000 + 20+ 1,000 + 60 + 1,000 + 10 (does not = 5,000)

Art of Jain, 1982 "The 12 Beings"

"The mathematical sciences
particularly exhibit order, symmetry, and limitation;
and these are the greatest forms of the beautiful."

Aristotle

AL-ARQUAN – AL – HINDU

"THE HINDU FIGURES"

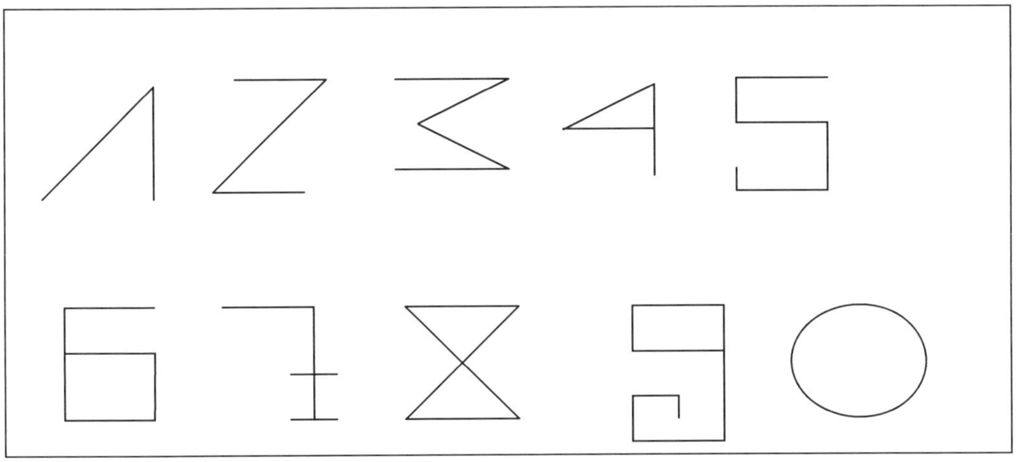

Have you ever wondered about the origin of the numbers we use today? where the actual hand-written script came from?
A lot of people call our Western use of Numbers as Arabic Numerals, but their origin goes further back to ancient Bharat or India!

EXERCISE 1

Mark with a coloured pencil or pen, with dots, the number of angles in each of the above 9 Hindu Numerals.

Notice on the above diagram that the Number 1 has 1 angle! and the Number 2 has 2 angles! Now you know why the 7 has a bar on its up-right. There are 7 angles.
It appears that the original numerical Sanskrit language was very angular and later in time these stark straight lines and angles became stylized into the current curving figures that we now use.

Below is perhaps the geometric origin of all numerals :
the 'URDHVATIRYAK SUTRA translated as :

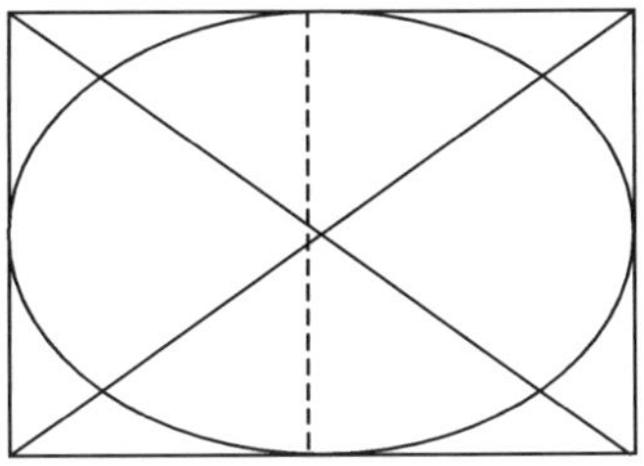

You can see here the origin of the sutra:
VERTICALLY AND CROSSWISE

Sub-SUTRA: BY THE COMPLETION or NON-COMPLETION

(Sanskrit title: Purana'purana'bhyam)

FOR ADDITION OF 2 DIGIT NUMBERS:

47 + 48

Best way to describe this to a student is to get them to visualize two pot of marbles, one pot has 47 and the other has 48. We must take or steal two marbles from the Pot of 47 to add to the Pot of 48 to make it 50....

Mentally, you can add these two digits quickly by adding "2" to the Pot "48". But we needed to borrow this "2" from the Pot "47" to make Pot 48 to become Pot 50 which is a round number.
This means we must reduce Pot 47 by 2, which now becomes:
"47 - 2" = 45.
Thus the two Pots now have 45 and 50 which means the addition is now **45 + 50**
which is easier to be adding "45 + 50" than it is to be adding 47 + 48.
The answer now becomes very obvious = **95**.

This process is something that most children naturally do, when adding their digits, it is a natural and simple process, and to give it a name, it is referred to as one of the **14 Sub-Sutras: BY THE COMPLETION**.

EXERCISE 2

Add the following 2 digit numbers, employing this Sub-Sutra: By The Completion.

a) 16 + 17 =

b) 27 + 28 =

c) 39 + 38 =

d) 105 + 117 =

e) 998 + 113 =

f) $\$14.99^{c} + \$7.23^{c}$

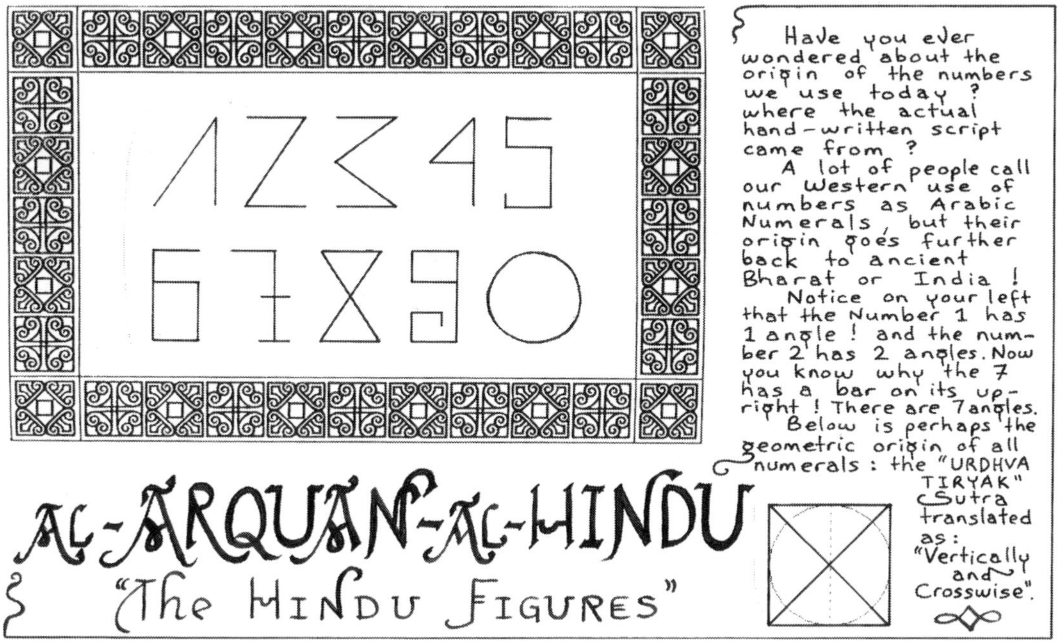

ANSWERS 2

a) 16 + 17 = (16-3)+(17+3) = 13+20 = 33
b) 27 + 28 = (27-2)+(28+2) = 25+30 = 55
c) 39 + 38 = (39+1)+(38-1) = 40+37 = 77
d) 105 + 117 = (105-3)+(117+3) = 102+120 = 222
e) 998 + 113 = (998+2)+(113-2) = 1000+111 = 1,111
f) $14.99 ^c^ + $7.23 ^c^ = 1499 + 723 = 1500 + 722 = 2000 + 222
= 2,222 =$22.22^c^

FOR ADDITION OF LARGER NUMBERS:

1	2	**3**
1	6	5
9	8	**7**
8	3	6
----	----	----
C3	C2	C1

Always look for the Pairs of 10, like 3+7=10 in the units column.

Call the 3 Columns: C1, C2 and C3

In C1, notice that the 3 and 7 have been bolded.
When adding the 6 and 5, think of it as 6 + 4 + 1, giving an immediate pair of 10 in the 6 + 4 and leaving a remainder of 1.
Thus C1 = two pairs of 10, giving 20 plus the 1 = 21.
Put the 1 down in the unit's column, and carry over the 2.

In C2, you have **2**+6+**8**+3+rem2 = 10+6+3+2 = 10+10+1 = 21.
Put the 1 down in the ten's column, and carry over the 2.

In C3, you have 1+**1+9**+8+rem2 = 1+10+**8+2** = 1+10+10= 21.

Thus the answer is 2,111.

EXERCISE 3

Add the following numbers utilizing "By The Completion".

a)

```
4 3 4 5
7 8 0 1
3 7 6 5
1 2 9 9
5 0 1 2
_______
```

b)

```
6 2 3
2 9 6
7 8 7
8 1 4
4 4 5
3 6 8
_____
```

REPRESENTING NUMBERS USING ONLY BASE 5 ie: The DIGITS From 1 To 5

Two thousand years ago, the Indian sages did not like working with numbers larger than 5, as it may have led to errors.
So they invented an ingenious system where all larger numbers of 6, 7, 8 and 9 were reduced or formatted to smaller numbers.
eg: the number 39 can be rewritten like this:

Visualize 39 as 40 less 1.
The Indians invented the concept of negative or minus numbers, and showed this by a bar or horizontal line above the negative number.

eg: 39 = 4 bar 1

It is written as 4**1** (with the 1 underlined)
(where the underlined "1" really has the flat bar above the number).

EXERCISE 4

Represent the following numbers in Base 5 utilizing the concept of Negative Numbers:

a) 19 **b)** 28 **c)** 67 **d)** 197 **e)** 117

This is **The Law of Economy** or **The Path of Least Resistance**, enabling Rapid Mental Calculation.

ANSWERS 3

a) 22,222 b) 3,333

ANSWERS 4

a) $2\underline{1}$ b) $3\underline{2}$ c) $7\underline{3}$ d) $20\underline{3}$ e) $12\underline{3}$

Ancient formula used for maths

By CLEO SRIBER,
Star correspondent

BY the application of a simple formula, a child can know that the cube root of 35,937 is 33 — without a calculator!

So says Jain, a mathematician from Mullumbimby, the author of seven self-published books on sacred geometry and Vedic mathematics, the culmination of 20 years of research.

Jain works with gifted and talented children and lectures at maths camps, universities and maths conferences.

As an authorised school performer, he has taught thousands of children in Australia through his travelling show, Mathemagics, teaching the wonder of pure mathematical principles.

Currently he is writing a series of four books for the Australian curriculum on Vedic mathematics..

Jain, author and mathematician.

"By relying on modern electronic calculators for the next 20 years or so, without cultivating mental calculations which build the 'mental muscle', we will sadly produce a generation of young adults whose brains will have atrophied," Jain says.

The brain is a muscle, he said, and it needs exercising.

Complex multiplications can be worked out in seconds, even by small children, such as converting the fraction 1/19 into a decimal form.

The answer is an 18-digit figure which requires 35 lines of exhausting long division to work out.

History is full of architectural marvels and mathematical wonders, that were calculated without calculators, or even pen and paper.

Following an ancient system of mathematics, rediscovered by an Indian mathematics and Sanskrit philosopher named Bharati Krsna Tirthaji, it is possible to solve every known mathematical problem, said Jain.

Jain will hold a workshop on Vedic mathematics in Lismore, with further information available on 6680 5833.

(taken from the Northern Star Newspaper 20-11-2002)

(the dancing digits)

SUTRA: BY THE DEFICIENCY

BASE UNDER TEN   CROSS-SUBTRACTION

We will now apply our understanding of ALL FROM NINE AND THE LAST FROM 10 to calculate the Multiplication of **SINGLE DIGITS BY SINGLE DIGITS**.

Obviously, all these numbers will be under 10, so the subtitle for this exercise is: Base Under 10

EXAMPLE 9 x 7

For 9 x 7 we say ALL FROM NINE AND LAST FROM TEN. When we only have single digits, like 9 x 7, the "All From Nine" part of the Sutra does not apply, only the "Last from Ten".

Which is really an understanding of subtracting both the 9 and the 7 from 10.

Mentally we are saying 9 is 1 less than 10 which is minus 1 or "-1".

And 7 is 3 less than 10 which is minus 3 or "-3".

The setting out will look like this:

To remember that you are using **Base Ten**, (as distinct from Base 20 or 100) have it written in a circle above the setting out, or in brackets, like this: (10)

(10) Always have your base written above or circled.

```
9 x 7 =  9 - 1    "Cross-Subtracting"
         7 - 3
        -------
      =  6 / 3
```

Notice, to get the "3" on the Right Hand Side, we multiply vertically the two Deficiencies, which is minus 1 times minus 3 giving plus or positive 3. (In my next book: Part 3, Multiplication), you will become familiar with another Sutra called: "Vertically and Crosswise", so be aware that you are multiplying the two deficiencies that are set out in a vertical alignment.

The answer has two parts, a left hand part and a right hand part. To clearly distinguish these two parts, the Hindu Mathematicians inserted a diagonal stroke or "forward slash" (/) underneath the minus signs, as shown above.

The minus signs indicates that the **Complements** are both less than Ten.
To get the "6" on the Left Hand Side, we perform a **CROSS-SUBTRACTION**, saying 9 minus 3 is the same as 7 minus 1. In fact, these two cross subtractions, as you will see, are always the same.
In fact, there are 4 ways of arriving at the Left Hand Side digit in the 10's column:

1. Add the two Ten's Digits (9 + 7) and subtract from 10 = 16 – 10 = **6**
2. Add the Deficiencies (1 + 3) and subtract from 10 = 10 – 4 = **6**
3. Cross-Subtract = 9 – 3 = **6**
4. Cross-Subtract = 7 – 1 = **6**

The Solution: 6 / 3 really means 6 lots of the base 10 which is 60, and 3 lots of the Units which is 3,
giving the answer of 60 + 3 = 63

When performing the mental calculation, Remember to **Cross-Subtract**: mentally saying: The diagonal relationship of the
(9 – 3) is always the same as the other diagonal (7 –1) both giving a sum of 6.

9 -1
X
7 -3

The reason the "**X**" is in bold is to highlight the fact that this interesting diagonal relationship of numbers relating to their deficiencies from 10 or 100 or 1000 etc was the origin or became the reason why we adopted the use of "**X**" as our symbol for Multiplication. Herein lies the desire for the brain to perform this even quicker, so you are encouraged to make even quicker mental short-cuts. Here is how an adept would multiply 9 x 7, with the least possible verbiage:
Nine less 3 equals 6, which is 60, and add to this one times three, equals 60 and 3, equals 63.
An even quicker mental short-cut would go like this, deleting any mental clutter:
Nine less three, six. Sixty. One times three. 3. Sixty-three.

This is the **PATH OF LEAST RESISTANCE**. It is also the **LAW OF ECONOMY**. This system will eventually be made global as it is highly efficient and stimulates the brain. Mental Calculation is superior to machine calculation. The western world adores **short-cuts**.

EXERCISE 5

Perform the following mental, One–Line Calculations for your common Times Table. Then write it out in the format shown above.
Multiply:

a) 8 x 7 **b)** 9 x 6 **c)** 8 x 8 **d)** 8 x 6 **e)** 9 X 9

INTRODUCING THE "CARRY-OVER"

Sometimes the number in the Units Column, derived from the multiplication of the two excesses, exceeds 10, at which point, we slide over the "1" similar to the western method.

EXAMPLE 7 X 6

Eg: 7 x 6 = 7 -3
= 6 -4

= 3 / $_{1}2$
= 4 / 2

The "1" of the "12" exceeds 10, so we carry over that "1" to the Tens Column, making 3 + 1 = 4
(Later on, when we get to Base 1**00** and Base 1**000** you will understand that according to the number of Zeroes "0" in our Base, will determine how many digits we can have on the Right Hand Side of the Forward Slash (**/**). In this case, we have one zero in our Base 1**0**, which indicates that we are forced to have only one digit on the Right hand Side. Base 100 allows 2 zeroes and Base 1,000 allows 3 zeroes on the Right hand Side).

EXERCISE 6

Perform the following mental, One–Line Calculations for your common Times Table. Then write it out in the format shown above.
Multiply:

a) 7 x 5 **b)** 6 x 6 **c)** 8 x 4 **d)** 7 x 3

ANSWERS 5&6

8 x 7. Mentally, I subtract from 8, the deficiency of 7 from 10, which is 8 – 3 which is 5 or really 5 lots of the Base 10 which is 50, and add to this the sum of the two deficiencies multiplied, which is –2 x –3 = 6, giving 50 + 6 = 56.
The working out would look like this: (see far right hand side).
Remember to Cross-Subtract: mentally saying: The diagonal relationship of the (8 – 3) is always the same as the other diagonal (7 – 2) both giving a sum of 5.

	(10)
8 x 7 =	8 – 2
	7 – 3

=	5 / 6

(10) **7 x 5** = 7 – 3 = 5 – 5 -------- = 2 / $_{1}$5 = 3 / 5 = **35**	(10) **6 x 6** = 6 – 4 = 6 – 4 ------- = 2 / $_{1}$6 = 3 / 6 = **36**	(10) **8 x 4** = 8 – 2 = 4 – 6 ------- = 2 / $_{1}$2 = 3 / 2 = **32**	(10) **7 x 3** = 7 – 3 = 3 – 7 ------- = 0 / $_{2}$1 = 2 / 1 = **21**

"I see a Future
where all Children of the World
are doing their Mathematics
swiftly in their Head"
(Jain 108)

ADDITION TRICK
ADD THESE 8 SIMPLE NUMBERS

Can You Add these following 8 Numbers, as fast as you see them.
They are easy numbers to add, but the speed of adding them
is what is being tested here.

1,000
10
1,000
20
1,000
60
1,000
10

(Answer is 4,100 not 5,000 !!!) The only people who get this correct, are those people who work everyday with numbers like bricklayers, carpenters, accountants, etc

DAY 4

- Sutra: **Trachtenberg's Method**: **2 Digit x 1 Digit** eg: 24x6
- Introducing all the **Cubic Numbers** eg: 6x6x6 = 36x6 =216

Sub-Sutra: TRACHTENBERG'S METHOD TO SOLVE 2 DIGITS X 1 DIGIT

Mentally solving 2 Digits x 1 Digit

In the realm of Multiplication, where any 2 Digit Numbers are multiplied by any single Digit Numbers, there is a beautiful and simple technique.

This system of multiplying 2 Digits by 1 Digit actually does not appear in Vedic Mathematics by Bharati Krsna Tirthaji. It first appeared to me in a book called:

"**The Trachtenberg Speed System of Basic Mathematics**"

but I have modified it to make it even simpler.

(Do Research on him: how this Swiss mathematician developed this system when a prisoner in a German Nazi camps).

The definition of Vedic Mathematics is 'Mental, One-Line Arithmetic'. I encourage all children to learn this amazingly swift mental method... moving across the line, rather than calculating traditionally down the page.

EXAMPLE How to mentally multiply 24 x 6 in a few seconds

Visualize "24" as "2" and "4" and multiplying both digits independently by "6" **.**

ie: Multiply by 6 both digits of 24. Thus upon your mental screen visualize

2 x 6 **/** 4 x 6 which becomes: 12 **/** 24.

(note: The forward slash **/** is used as a place marker only).

Sometimes, it can also be written as: 12 **/** $_2$4.

Your focus now is to have only one digit on the right-hand-side of the slash.

The quick answer is achieved by adding the "2" of the "24" over to the left-hand-side, appearing like this:

12 + **2** / 4 = 14 / 4.

When done quickly in the Mind, the mental steps are:

24 x 6 = 2 x 6 / 4 x 6 = 12 / 24 = 12+2 / 4 = 14 / 4 = **144.**

EXAMPLE 44 x 6 Using DIGITAL SUMS

44 x 6 = 4x6 / 4x6 = 24 / 24 = 24+2 / 4 = 26 / 4 = **264.**
It's that simple, keep practicing with other examples:

EXAMPLE 18 x 8 Using DIGITAL SUMS

18 x 8 = 1x8 / 8x8 = 8 / 64 = 8+6 / 4 = 14 / 4 = **144.**

EXAMPLE 71 x 7 Using DIGITAL SUMS

71 x 7 = 7x7 / 7x1 = 49 / 7 = **4 9 7.**
(Notice there is no digit to slide over to the left).

EXERCISE 1

Mentally Multiply the following:

a) 23 x 6 =

b) 48 x 4 =

c) 52 x 8 =

d) 64 x 7 =

e) 75 x 5 =

f) 88 x 8 =

g) 61 x 6 =

h) 97 x 7 =

i) 134 x 6 =

j) 243 x 4 =

k) 456 x 7 =

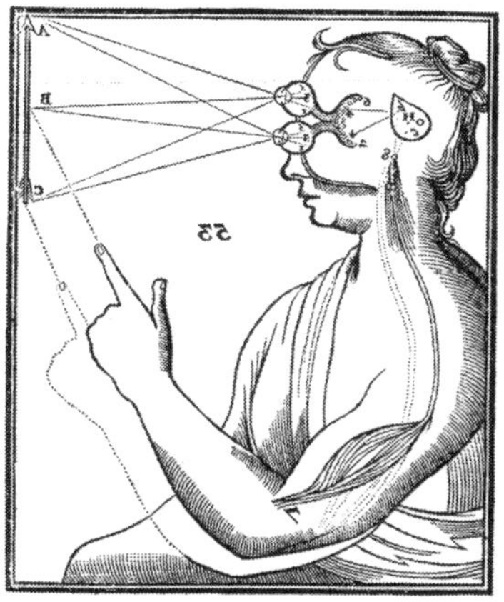

TRATAKA: **Mindless Gazing**: To strengthen the 3rd Eye.

The definition of Vedic Mathematics is 'Mental, One-Line Arithmetic'. I encourage all children to learn this amazingly swift mental method... moving across the line, rather than calculating traditionally down the page. It is of great value to do such sums upon the visualized 'blackboard in your mind'. The ancient Indian civilization practiced mental gazing techniques, called 'Trataka', where the fixed gaze ultimately awakened the Third Eye and gave insight into the science of Complementary Colors in the Artist's Color Wheel, e.g. staring at a red star for 3 minutes, then closing the eyes produces an after effect of a green star! Thus, 'holding' these numbers in your Mind's Eye, helps build the Mental Muscle. Vedic Mathematics is really about re-learning the lost art of 'Holding Digits on the Inner Mental Screen'.

When in the movie called the 'Rain Man' the central character or Idiot Savante was asked: "How do you perform correctly such complex mathematical tasks"? He replied: "**I see it**".

This is the **supreme Art of Pattern Recognition** symbolized by the Right Visual Cortex of the Feminine Intuitive side of the Brain. Enjoy this ancient code.

ANSWERS 1

a) 23 x 6 = 2 x 6 / 3 x 6 = 12 / 18 = 13 / 8 = 138

b) 48 x 4 = 4 x 4 / 8 x 4 = 16 / 32 = 19 / 2 = 192

c) 52 x 8 = 5 x 8 / 2 x 8 = 40 / 16 = 41 / 6 = 416

d) 64 x 7 = 6 x 7 / 4 x 7 = 42 / 28 = 44 / 8 = 448

e) 75 x 5 = 7 x 5 / 5 x 5 = 35 / 25 = 37 / 5 = 375

f) 88 x 8 = 8 x 8 / 8 x 8 = 64 / 64 = 70 / 4 = 704

g) 61 x 6 = 6 x 6 / 1 x 6 = 36 / 06 = 36 / 6 = 366

h) 97 x 7 = 9 x 7 / 7 x 7 = 63 / 49 = 67 / 9 = 679

i) 134 x 6 = 1 x 6 / 3 x 6 / 4 x 6 = 6 / 18 / 24 = 6 / 18 + 2 / 4
= 6 / 20 / 4 = 6 + 2 / 0 / 4 = 8 / 0 / 4 = 804

j) 243 x 4 = 2 x 4 / 4 x 4 / 3 x 4 = 8 / 16 / 12 = 8 / 16 + 1 / 2
= 8 / 17 / 2 = 8 + 1 / 7 / 2 = 9/7/2 = 972

k) 456 x 7 = 4 x 7 / 5 x 7 / 6 x 7 = 28 / 35 / 42 = 28 / 35 + 4 / 2
= 28 / 39 / 2 = 28 + 3 / 9 / 2 = 31/9/2 = 3,192

TRACHTENBERG METHOD
TO SOLVE THE SEQUENCE OF CUBIC NUMBERS

At this stage, we are going to do another set of exercises, the same again with 2D x 1D (short for 2 Digits x 1 Digit) to determine the Sequence of CUBIC NUMBERS. It will be necessary to remember these Cubic Numbers, as you will be required to learn how to mentally solve Cubic Roots achieved by mentally remembering this Sequence of Cubic Numbers: 1, 8, 27, 64, 125, 216, 343, 512, 729 ...

EXAMPLE 6x6x6

6x6x6 is abbreviated to "**6 Cubed**" or "**6 To The Power of Three**" or "$\mathbf{6^3}$". This Power of 3 is also known as an "**Index**" whose plural form is "**Indices**". Another synonym or word that means the same thing is: "**Exponent**".

To rewrite 6x6x6 in the Trachtenberg form of 2D x 1D we write the 6x6 part as 36 and multiply this by 6.

Thus 6x6x6 = 36x6

Apply the principle of multiplying the 36 by 6 expressed as two pairs: 36x6 = 18 / 36 but the rule is we are only allowed one digit on the right hand side of the forward slash, so we express 6x6x6 as: 18 / $_3$6 then slide the "baby 3" or "carry-over" to the left hand side, giving the final answer as 21/6 or **216**.

As a Mental, One-Line Answer "$\mathbf{6^3}$" can be written in this format:
$\mathbf{6^3}$ = 6x6x6 = 36x6 = 18 / $_3$6 = 21/6 = **216**.

EXERCISE 2

Give the answers to the following Cubic Numbers mentally, then write out the One-Line Answer format as shown above:

a) $\mathbf{4^3}$ =

b) $\mathbf{5^3}$ =

c) $\mathbf{7^3}$ =

d) $\mathbf{8^3}$ =

e) $\mathbf{9^3}$ =

nb: In this exercise, you will see that you will have to know your Times Tables, testing your knowledge of "Magic Fingers" to calculate the Series of Squared Numbers like 7^2 aka 7x7.

ANSWERS 2

a) $\mathbf{4^3}$ = 4x4x4 = 16x4 = 4 / $_2$4 = 6/4 = **64**.

b) $\mathbf{5^3}$ = 5x5x5 = 25x5 = 10 / $_2$5 = 12/5 = **125**.

c) $\mathbf{7^3}$ = 7x7x7 = 49x7 = 28 / $_6$3 = 34/3 = **343**.

d) $\mathbf{8^3}$ = 8x8x8 = 64x8 = 48 / $_3$2 = 51/2 = **512**.

e) $\mathbf{9^3}$ = 9x9x9 = 81x9 = 72 / 9 = **729**.

(There is an amazing chapter called: "**COMPUTING BY BRAIN**" on pages 175 to 178, in this book "The Life and Science of Richard Feynman" by James Gleick, author of "Chaos", highlighting how the atomic-bomb builders, in the 1940s could calculate quicker than any computer, and this was prior to Bharati Krishna Tirthaji's arrival to Los Angeles in 1952! It goes to show that these brilliant mathematical sutras or formulas belong to the Universal Mind and are not just a Vedic phenomenon). Towards a Universal Mathematics Curriculum embracing all Cultures.

DAY 5

- **Multiplication of the Teen Numbers**
- **Sutra: "By The Excess"** for "Over Base 10" having "**No Carry-Over**" eg: 12x13 then "**Carry-Overs**" eg: 13x14
- Sub-Sutra: "**On The Flag**"
- **Squaring of the Teen Numbers**

MULTIPLICATION OF THE TEEN NUMBERS FOR TEENAGERS

Sutra: "By The Excess"
(Multiplication of Numbers over a Base)
(Base Over Ten) (Cross Addition)

Here is the trick how to multiply one teen number by another teen number, in your head, in seconds!

TEEN NUMBERS WITH NO CARRY-OVERS

By applying the formula or Sutra: "By The Excess", you can instantly solve the multiplication of any Teen Numbers.

EXAMPLE 12 x 13

Just think of the two excesses, asking yourself, how much are these two numbers more than the Base of 10, and raise those numbers on your **INNER MENTAL SCREEN**. The excesses are 2 and 3. All you need to do is add the excess of 2 to the other number 13, or add the excess of 3 to the other number 12, and in both cases the sum is 15 or 15 lots of the base 10, at which point we hold in our Memory Bank the number of 15 x 10 = 150, and add to this the multiplication of the two excesses which is 2 x 3 = 6, giving the mental, one-line answer of 150 + 6 = 156.

12 x 13 = 12 + the excess of 3 / multiply the two excesses of 2 x 3
= 12 + 3 / 2 x 3
= 15 / 6
= 156

Another way of simplifying these Multiplications of the Teen Numbers, is to visualize the Future Answer in your Mental Screen or Window, and say that "The Answer is Something, Forward Slash, and One Digit allowed to the right hand side".

The picture or structure for the answer to descend into has this appearance:

____________________ / ____

which is expressed as "Something, with One Digit to the Right".
This allows the Right Feminine Brain, or Visual Cortex or picture-side, to kick in and see the instant answer.

Vedic Mathematics has a subtitle called "Mental, One-Line Arithmetic,
which means that if we can not calculate something instantly in our head then we can express it as a One-Line Answer.
Here is the simplified format which will be required for the following exercises in this chapter:

12 X 13 = 12+3 / 2x3 = 15/6 = 156

EXERCISE 1

Multiply the following Teen Numbers mentally, which have no "carry-overs", then show the setting out on paper:

a) 12 x 14 =

b) 19 x 11 =

c) 13 x 13 =

d) 11 x 15 =

e) 12 x 12 =

ANSWERS 1

a) 12 x 14 = 12 + 4 / 2 x 4 = 16 / 8 = 168
b) 19 x 11 = 19 + 1 / 9 x 1 = 20 / 9 = 209
c) 13 x 13 = 13 + 3 / 3^2 = 16 / 9 = 169
d) 11 x 15 = 11 + 5 / 1 x 5 = 16 / 5 = 165
e) 12 x 12 = 12 + 2 / 2 x 2 = 14 / 4 = 144

TEEN NUMBERS WITH CARRY-OVERS

Similarly, by applying the same formula, you can instantly solve say **13 x 17** by use of this Sutra: **By The Excess**.
Just think of the two excesses, asking yourself, how much are these two numbers more than the Base of 10, and raise those numbers on your **INNER MENTAL SCREEN**. The excesses are 3 and 7. All you need to do is add the excess of 3 to the other number 17, or add the excess of 7 to the other number 13, and in both cases the sum is 20 or 20 lots of the base 10, at which point we hold in our Memory Bank the number of 20 x 10 = 200, and add to this the multiplication of the two excesses which is 3 x 7 = 21, giving the mental, one-line answer of 221.

13 x 17 = 17 + the excess of 3 / multiply the two excesses of 3 x 7
= 17 + 3 / 3 x 7
= 20 / 21 but we are only allowed "One Digit To The Right" because our Base 10 has only One Zero, so we need to rewrite the right hand side highlighting the "carry-over" which gets added or slid over to the left hand side.
= 20 / $_{2}1$
= 22/1
= 221

It will be required in Exercise 2, to give the One-Line Answers in this format:

13 x 17 = 13+7 / 3x7 = 20 / $_{2}1$ = 22/1 = 221

EXERCISE 2

Multiply the following Teen Numbers mentally, which have "carry-overs", then show on paper the setting out as One-Line Answers:

a) 13 x 14 =

b) 12 x 16 =

c) 15 x 16 =

d) 17 x 18 =

e) 19 x 12 =

f) 13 x 16 =

g) 14 x 17 =

h) 16 x 14 =

ANSWERS 2

a) 13 x 14 = 13+4/3x4 = 17/12 = 17/$_1$2 = 18/2 = 182
b) 12 x 16 = 12+6/2x6 = 18/12 = 18/$_1$2 = 19/2 = 192
c) 15 x 16 = 15+6/5x6 = 21/30 = 21/$_3$0 = 24/0 = 240
d) 17 x 18 = 17+8/7x8 = 25/56 = 25/$_5$6 = 30/6 = 306
e) 19 x 12 = 19+2/9x2 = 21/18 = 21/$_1$8 = 22/8 = 228
f) 13 x 16 = 13+6/3x6 = 19/18 = 19/$_1$8 = 20/8 = 208
g) 14 x 17 = 14+7/4x7 = 21/28 = 21/$_2$8 = 23/8 = 238
h) 16 x 14 = 16+4/6x4 = 20/24 = 20/$_2$4 = 22/4 = 224

THE SQUARING OF THE TEEN NUMBERS

This is related to the 7th Sub-Sutra concerning "Deficiencies"
but in this instance it relates to: **BY THE EXCESS**,
or cryptic word-formula:

"By whatever the extent of the excess, add on that excess and square that excess".

Often these formulas appear to make no sense, but when they are applied, they make the calculations effortless. Simply add on the excess and square it. You are mainly concerned here with that excess number above ten. Eg when I am squaring fourteen ie: 14 x 14, then the excess is "4". I am mentally flagging that "4" in my consciousness or raising that "4" upon my Inner Mental Screen. Its that simple, add it and square it.

EXAMPLE 14 X 14

$14 \times 14 \quad = 14 + \mathbf{4} / \mathbf{4}^2 = 18 / {}_{1}6 = 19 / 6 = 196$

nb: (this involved a "carry-over").
The Mental Path of Least Resistance would quickly say:
14, 4, 180, $_{1}6$, 196. This is the swiftest level of calculation that you are striving for.
The setting out for this example would look like this:
This is the long way, it is only shown for theory, so keep doing the mental calculations first.

```
14 x 14   = 14 + 4
   |        14 + 4
            -------
          = 18 / ₁6

          = 19 / 6
          = 196
```

Observe that the cross addition of

14 + 4
14 + 4

which has a sum of 18, involves the above 2 lines of setting out, can be smartly shortened to a mental one-line answer,
when you say: 14 + the excess of the other number, which is 4, makes 18.
It is actually 18 lots of the Base Ten, so straight away you can say 180, which gives the Left Hand Side of the Answer; then I add to 180 the excesses 4 x 4 which is 16, making a final total of 196.

EXERCISE 3

Square the following Teen Numbers:

a) **11** x **11 =**

b) **12** x **12 =**

c) **13** x **13 =**

d) **15** x **15 =**

e) **16** x **16 =**

f) **17** x **17 =**

g) **18** x **18 =**

h) **19** x **19 =**

nb: It is good practice, to learn this Sequence of Squared Numbers, known algebraically as "$\mathbf{n^2}$".
This helps build your **Memory Power** and **Mathematical Confidence.**

ANSWERS 3

a)	11 x 11	$= 11 + 1 / 1^2$	= 12 / 1	= 121	
b)	12 x 12	$= 12 + 2 / 2^2$	= 12 / 4	= 144	
c)	13 x 13	$= 13 + 3 / 3^2$	= 16 / 9	= 169	
d)	15 x 15	$= 15 + 5 / 5^2$	= 20 / 25	= 22/5	= 225
e)	16 x 16	$= 16 + 6 / 6^2$	= 22 / 36	= 25/6	= 256
f)	17 x 17	$= 17 + 7 / 7^2$	= 24 / 49	= 28/9	= 289
g)	18 x 18	$= 18 + 8 / 8^2$	= 26 / 64	= 32/4	= 324
h)	19 x 19	$= 19 + 9 / 9^2$	= 28 / 81	= 36/1	= 361

Sub-SUTRA: ON THE FLAG

2 3

| |

12 x 13 = ________ / __

The section where we hoist or raise the excess of the teen numbers above 10, is a sub-sutra known as **ON THE FLAG**.
As soon as we have created this structure with On The Flag,
By merely raising the 2 and the 3,
the student is able to instantly see the answer because we have created the necessary structure for the answer to appear in.
It's a bit like in the introduction to the Matrix movie where we see this Digital Rain or cascading of numbers falling down the screen. This is thus a critical part in the ability to forsee the answer. This is Right Brain Feminine Mathematics that induces Whole Brain Learning. It's the future in how we will calculate.

At this stage, we need to highlight or tick or mark "done" this relevant sub-sutra shown on our chart on page 19.

DAY 6

- **Revision** of the Previous Week.
- Nikhilam **Base 20**
  Using the Sutra: **By The Deficiency** to solve calculations like 18 x 19.
  Using Sub-Sutra: "**The Remainder Remains The Same**".
- Nikhilam **Base 20** using the Sutra: **By The Excess** to solve calculations like 23 x 24
- Nikhilam **Base 30** using the Sutra: By The Excess to solve calculations like 33 x34
- **Squaring of the Numbers from 50 to 59** eg: 52^2
- **Base 90**: By The Deficiency, Under the Base 100 to solve 98x97
  Using Sub-Sutra: "**Lessen By The Deficiency**"
- Nikhilam **Base 100** using the Sutra: By The Deficiency to solve **The Squaring of the Numbers in the Nineties** like 93^2
- **Base 100**: **By The Excess**, Over the Base to solve numbers from 100 to 120 eg: 108x109 and 113 x 114
- **Base 1,000** and higher eg: 1011x1023

(Jain's Logo: Magic Square of 3x3 drawn 4 Times at 45 degrees)

"**omnia apud me mathematica fiunt**:
With me everything turns into mathematics."
Descartes

HOW NEW SUTRAS ARISE: SHIFTING FROM BASE 10 TO BASE 20

Multiplying the Higher Teen Numbers Using Base Twenty and Using the Sutra: By The Deficiency + Cross-Subtraction

One reason for giving you the progressively increasing multiplications of the Teen Numbers, is to demonstrate that they are easy from around 13 to 14, but get more difficult when you get to 17, 18 and 19. The second last example, 18 x 19, actually gets clumsy, so the yogis, from several thousand years ago, each time a problem arose, solved it with better short-cuts. So the example given here is to appreciate that solving 18 x 19 by using the Nikhilam Sutra: **By The Excess**, is clumsy, the methodology breaks down, and it is more efficient to step up to the next base: **BASE 20** and utilize the Nikhilam Sutra: **By The Deficiency**.

EXAMPLE 18 X 19

18 x 19 Our Original Base is 10 but our Working Base is 20.
We must ask ourselves, how much are these two numbers less than 20?
Here is the setting out.

(20)	
18 − 2	18 is deficient from the Working Base 20 by 2
19 − 1	19 is deficient from the Working Base 20 by 1
-------	The 2 negative signs (-2 x −1) cancel out, according to the rule: minus x minus makes a positive number.
17 / 2	
x2	We must multiply the 17 lots of Base 10 by 2. NB: **The Remainder remains Constant**, ie: it is not doubled.
34 / 2	
= 342	

The preferred Mental, One-Line Solution goes like this:
From 18 we subtract the deficiency of the other number which is (20 - 19) making 18 less 1 which is 17, we double that as we are in Base 20, making 34 or 34 lots of my original Base 10, (being 340) and add on to this the multiplication of those two deficiencies which is 1 x 2 = 2, giving 340 + 2 = 342.
This is summarized below in the standard
One-Line Written Answer form required for the exercises:

18 x 19 = (18 - 1) x 2 / 1 x 2 = 17 x 2 / 2 = 34/2 = 342

EXERCISE 1

Multiply the following Teen Numbers, using a working Base of 20 and applying the necessary and most simplest formula:
By The Deficiency. Do it Mentally, as a One-Line Answer, then write it down as shown in the above setting out.

a) 17 x 18 =

b) 18 x 16 =

c) 19 x 19 =

d) 16 x 19 =

e) 16 x 17 =

ANSWERS 1

a) 17 x 18 = (17 – 2) x 2 / -2 x –3 = 15 x 2 / 6 = 30/6 = 30
b) 18 x 16 = (18 – 4) x 2 / -2 x –4 = 14 x 2 / 8 = 28/8 = 288
c) 19 x 19 = (19 – 1) x 2 / -1 x –1 = 18 x 2 / 1 = 36/1 = 361
d) 16 x 19 = (16 – 1) x 2 / -4 x –1 = 15 x 2 / 4 = 30/4 = 304
e) 16 x 17 = (16 – 3) x 2 / -4 x –3 = 13 x 2 / $_1$2 = 26 /$_1$2 = 27/2 = 272

nb: This last example involved use of the Sutra:
"**The Remainder Remains The Same**".
Notice in the part where it says:
= 13 x 2 / $_1$2 = 26 / $_1$2 means that the answer is a result of doubling the 13 to give 26 lots of the Base 10 which is 260, and add on to that the Remainder of 12. But if we were to carry over early the "1" of the "2" in "$_1$2" and add it on to the 13, giving 14, which must be doubled to 28, creates the wrong answer.

BASE 20 USING SUB – SUTRA: "BY THE EXCESS" MULTIPLYING NUMBERS IN THE TWENTIES

Defining the "Working Base" and the "Original Base"

Multiplying two numbers in the 20s is fairly simple, it just involves one more step, a **doubling** of the digits on the Left Hand Side.

EXAMPLE 23 X 24

What is **23 x 24** ? At this point, we need to clarify the difference between the **Original Base of 10** and the **Working Base of 20**. For all the following Bases from 30 to 80, our Original Base will always be Base 10, but our Working Base can change to be the nearest round number that is closest to both numbers being multiplied. Until we arrive to base 100, we are still governed to have only one number to the right of the forward slash symbol. The mental working out for 23 x 24 is: 23 + the excess of the other number which is 4, makes 23 + 4, giving 27. This number I double immediately. Double 27 is 54, which is really 54 lots of my Base 10, giving 540 to which I add the product of the two excesses which is 3 x 4 = 12. The answer is therefore 540 + 12 = 552.
To express this as a Mental One-Line Answer is as follows:
23 x 24 = (23 + 4) x 2 / 3 x 4 = 27 x 2 / 12 = 54 / $_1$2 = 55/2 = 552
To write it out another way shows the Cross-Addition visible, just for your viewing. Though, it is best not to imitate this style so that you are encouraged to acquire the habit of performing the calculations mentally:

{20}	Always place the Base you are in, either in brackets or a circle, at the top of the working out.
23 + 3	Read this like this: 23 exceeds the Base 20 by 3, and 24 exceeds the Base 20 by 4
24 + 4	nb: the Cross Addition says that the (23 + 4) is the same as the (24 + 3) both = 27
--------	
27 / 12 **x2**	nb: Only the Cross-Addition is doubled. The Multiplication of the Excesses, (12) remains the same.
--------	This was explained previously as a Sub-Sutra called: The Remainder Remains The Same.
54 / $_1$2	
55 / 2	Thus the answer is 55 lots of Base 10 plus 2 = 552

EXERCISE 2

Perform the following multiplication of numbers in the twenties:

a) 21 x 22 =

b) 25 x 26 =

c) 23 x 25 =

d) 26 x 28 =

e) 27 x 29 =

f) 21 x 26 =

g) 22 x 28 =

h) 25 x 25 =

ANSWERS 2

a) 21 x 22 = (21 + 2) x 2 / 1 x 2 = 23 x 2 / 2 = 46 / 2 = 462
b) 25 x 26 = (25 + 6) x 2 / 5 x 6 = 31 x 2 / 30 = 62 / ${}_{3}0$ = 65/0 = 650
c) 23 x 25 = (23 + 5) x 2 / 3 x 5 = 28 x 2 / 15 = 56 / ${}_{1}5$ = 57/5 = 575
d) 26 x 28 = (26 + 8) x 2 / 6 x8 = 34 x 2 / 48 = 68 / ${}_{4}8$ = 72/8 = 728
e) 27 x 29 = (27 + 9) x 2 / 7 x 9 = 36 x 2 / 63 = 72 / ${}_{6}3$ = 78/3 = 783
f) 21 x 26 = (21 + 6) x 2 / 1 x 6 = 27 x 2 / 06 = 54 / 6 = 546
g) 22 x 28 = (22 + 8) x 2 / 2 x 8 = 30 x 2 / 16 = 60 / ${}_{1}6$ = 61/6 = 616
h) 25 x 25 = (25 + 5) x 2 / 5 x 5 = 30 x 2 / 25 = 60 / ${}_{2}5$ = 62/5 = 625

BASE 30

Multiplying two numbers in the 30s is fairly simple. It just involves one more step, a **tripling** of the digits on the Left Hand Side.

EXAMPLE 33 X 34

What is **33 x 34**? The mental working out is: 33 + the excess of the other number which is 4, makes 33 + 4, giving 37. This number I triple immediately, as my base now, is not Base 10, but Base 30. Triple 37 is 111, which is really 111 lots of my Base 10, giving 1,110 to which I add the product of the two excesses which is 3 x 4 = 12. The answer is therefore 1,110 + 12 = 1,122.

To express this as a Mental One-Line Answer is as follows:
33 x 34 = (33 + 4) x 3 / 3 x 4 = 37 x 3 / 12 = 111 / $_{1}2$ = 112/2
= 1,122

To write it out another way shows the Cross-Addition visible, just for your viewing. Though, it is best not to imitate this style so that you are encouraged to acquire the habit of performing the calculations mentally:

(30)	Original Base = 10, Working Base = 30
33 + 3	Read this like this: 33 exceeds the Base 30 by 3, and
34 + 4	34 exceeds the Base 30 by 4 nb: the cross addition says that the (33 + 4) is the same as the (34 + 3), both = 37
---------	
37 / 12	
x3	nb: Only the Cross-Addition is tripled. The Multiplication of the Excesses, (3x4=12) remains the same.
---------	
111 / $_{1}2$	nb: We are only allowed one digit on the Right Hand Side, so the "1" of the "12" gets carried over.
112 / 2	Thus the Answer = 1,122

EXERCISE 3

Perform the following multiplication of numbers using the Working Base of Thirty:

a) 31 x 32 =

b) 35 x 36 =

c) 33 x 35 =

d) 36 x 38 =

e) 37 x 39 =

f) 31 x 36 =

g) 32 x 38 =

h) 35 x 35 =

i) 28 x 29 =

j) 27 x 28 =

ANSWERS 3

a) 31 x 32 = (31 + 2) x 3 / 1 x 2 = 33 x 3 / 2 = 99/2
= 992

b) 35 x 36 = (35 + 6) x 3 / 5 x 6 = 41 x 3 / 30
= 123 / $_{3}$0 = 126/0 = 1,260

c) 33 x 35 = (33 + 5) x 3 / 3 x 5 = 38 x 3 / 15
= 3x3/8x3 / $_{1}$5 = 9/24/ $_{1}$5 = 11/4/$_{1}$5 = 11/5/5=1,155

nb: for solving "38 x 3", remember in Chapter or Day 4, the Trachtenberg Method, we multiplied 2 Digits by 1 Digit, by considering "38" as "3" and "8" and multiplying both digits independently by 3 giving: 3x3/8x3 or 9/$_{2}$4 = 11/4 or 114

d) 36 x 38 = (36 + 8) x 3 / 6 x 8 = 44 x 3 / 48 = 12/12 / $_{4}$8
= 13/2 / $_{4}$8 = 13/6 /8 = 1,368

e) 37 x 39 = (37 + 9) x 3 / 7 x 9 = 46 x 3 / 63 = 12/18 / $_{6}$3
= 13/8 / $_{6}$3= 13/14 /3 = 14/4 /3 = 1,443

f) 31 x 36 = (31 + 6) x 3 / 1 x 6 = 37 x 3 / 06 = 9/21 / 6
= 111/6 = 1,116

g) 32 x 38 = (32 + 8) x 3 / 2 x 8 = 40 x 3 / 16 = 120 / $_{1}$6
= 121/6 = 1,216

h) 35 x 35 = (35 + 5) x 3 / 5 x 5 = 40 x 3 / 25 = 120 / $_{2}$5
= 122/5 = 1,225

i) 28 x 29 = (28 – 1) x 3 / -2 x –1 = 27 x 3 / 2 = 81/2
= 812

j) 27 x 28 = (27 – 2) x 3 / -3 x –2 = 25 x 3 / 6 = 75/6
= 756

THE SQUARING OF NUMBERS FROM 50 TO 59

Here is a simple, delightful and elegant way to square all the numbers in the 50s, mentally, in 3 quick steps:
(Technically, our Original Base is Base 50, but our Working Base is Base 100).

EXAMPLE 56^2

STEP 1: $\mathbf{56^2}$ = Visualize the predicted Base 100 format for the digits to appear in ________ / _ _
ie: "something with 2 digits after it"
The 2 digits on the Right are determined by the 2 Zeroes in the Base 1**00**.
Square the last digit and tag this to the end of the predicted format: = _______ / 3 6

STEP 2: Square the first digit "5" of the "**5**6" and add the last digit of the "5**6**", and there's your answer!

= $5^2 + 6$ / 36
= 25 + 6 / 36

STEP 3: = 31 / 36
= 3,136

As a Mental One-Line Answer, it would appear as:

5x5 + 6 / 6x6 = 31 / 36 = 3,136.

EXERCISE 4

Square the following numbers in the Fifties. Do them both mentally and in the format shown above as One-Line Answers.

a) 51 x 51 =

b) 52 x 52 =

c) 53 x 53 =

d) 54 x 54 =

e) 55 x 55 =

f) 57 x 57 =

g) 58 x 58 =

h) 59 x 59 =

The swiftness of arriving at these answers is quite dazzling.
When you have practiced **the Squaring of all the Numbers in the 50s** you will realize how simple it is.
Perhaps you will start using these particular examples as a fun and startling party trick.
The goal obviously is not to show off and pretend that you have mathematical occult powers, but rather that the rapidly forth-coming answers are based on pure patterns! And that they are shareable, not mysterious.

ANSWERS 4

a) 51 x 51 = 5x5 + 1 / 1 x 1 = 25 + 1 / _ 1 = 26 / 01 = 2,601
nb: there must be 2 digits on the Right Hand Side of the Forward Slash.
Numbers under 10 are prefixed by Zero in this form "01" or "04" or "09" as we are in Base 100.
b) 52 x 52 = 5x5 + 2 / 2 x 2 = 25 + 2 / _ 4 = 27 / 04 = 2,704
c) 53 x 53 = 5x5 + 3 / 3 x 3 = 25 + 3 / _ 9 = 28 / 09 = 2,809
d) 54 x 54 = 5x5 + 4 / 4 x 4 = 25 + 4 / 16 = 29 / 16 = 2,916
e) 55 x 55 = 5x5 + 5 / 5 x 5 = 25 + 5 / 25 = 30 / 25 = 3,025
f) 57 x 57 = 5x5 + 7 / 7 x 7 = 25 + 7 / 49 = 32 / 49 = 3,249
g) 58 x 58 = 5x5 + 8 / 8 x 8 = 25 + 8 / 64 = 33 / 64 = 3,364
h) 59 x 59 = 5x5 + 9 / 9 x 9 = 25 + 9 / 81 = 34 / 81 = 3,481

BASE 90
MULTIPLYING NUMBERS IN THE NINETIES

INTRODUCING BASE 100.
USING THE SUTRA: BY THE DEFICIENCY

EXAMPLE 92 x 93

Watch how it can be solved within 4 seconds of mental working out when performed in the correct Base of 100, and using the sutra: By The Deficiency

(**100**)	Always have the base circled at the top of your setting out
92 - 08 93 - 07 ---------	You can cross-subtract two ways: (92 - 07) or (93 - 08) Both cross-subtractions give 85 or 85 lots of the Base 100
85 / 56	
8,556	

As a Mental One-Line Answer: The mental chatter would go like this, ultimately simplified:
From 92, I subtract the Deficiency from 100 of the other number which is (100 - 93 = 7) making 92 - 7 = 85, which is really 85 lots of my Base 100, so I say 8,500. I now add to this the multiplication of the two Deficiencies which 8 x 7 = 56, giving me the quick answer of 8,500 and 56 = 8,556.

Quickly, I could just as well say:
92 less 7, 8500, 8 x 7, 56, 8,556.
You want to get to this level, no mucking around.

As a One-Line of Arithmetic:
92 x 93 = 92 – 7 / 8 x 7 = 85 / 56 = 8,556

Isn't that Beautiful! So fast, so efficient.

EXERCISE 5

Use both methods shown above: working down the page and working swiftly across the line, in the Base of 100, to solve the following:

a) 98 x 97 =

b) 96 x 95 =

c) 94 x 93 =

d) 98 x 92 =

e) 97 x 96 =

f) 98 x 94 =

g) 88 x 98 =

ANSWERS 5

a) 98 x 97	= 98 – 3 / -02 x -03	= 95 / 06	= 9,506
b) 96 x 95	= 96 – 5 / -04 x -05	= 91 / 20	= 9,120
c) 94 x 93	= 94 – 7 / -06 x -07	= 87 / 42	= 8,742
d) 98 x 92	= 98 – 8 / -02 x -08	= 90 / 16	= 9,016
e) 97 x 96	= 97 – 4 / -03 x -04	= 93 / 12	= 9,312
f) 98 x 94	= 98 – 6 / -02 x -06	= 92 / 12	= 9,212
g) 88 x 98	= 88 – 2 / -12 x -02	= 86 / 24	= 8,624

THE SQUARING OF THE NUMBERS IN THE NINETIES

"By whatever the extent of the Deficiency, lessen it still further by that Deficiency, and set up the Square thereof of that Deficiency".

EXAMPLE 99 x 99 or 99^2 in Base 100

99 Squared:
Think only of the Deficiency which is (100 – 99) = **1**.
The Sutra above says to lessen the Number 99 by that Deficiency which is 1,
making it (99 – 1) = 98 or 98 lots of the Base 100
which is really 9,800,
and add to that the square of the Deficiency
which is (1 x 1) or 1^2 **=** 1.
So the answer is 9,800 + 1 = 9,801.

As a swift One-Line Answer:

99 Squared: = 99 – 1 / 1 x 1
= 98 / **0**1
= 9,8**0**1

THE INVENTION OF ZERO

Remember that we need 2 Digits on the Right Hand Side of the Forward Slash governed by the 2 Zeroes in our Base 100.
The reason that the Zero is highlighted above, is that here is the **Origin of the humble Zero**, as we shifted from Base 10 to Base 100, our answers evolved intelligently to accommodate a highly sophisticated Place Value System, that led also to the discovery of the Decimal System and the Decimal Point that we all now use. It was used by the Hindus, several thousand of years ago, with rumors of having its origin in Iraq, ancient Babylonia 12,000 years ago, that today amazes the modern world with its ingenious use of its stepped pyramids or ziggurats that were really Planetary Observatories, the mathematics of which could not have been achieved without use of the Zero.
I believe now that the greatest invention in 2,000 years was the origin or Rememberance of and the use of the Zero, for without it we would have no binary code, therefore no modern silicon-chipped computers, therefore no rockets to the moon, in fact, no modern civilization as we know it. Without this quaint, little, shy and modest Zero, we would still be in the Dark Ages.
Ultimately, this history and study of mathematics is a legacy of many ancient cultures, we just can't say that it all belongs to the "Vedas" which were "written on the Rays of the Sun", yes they were, but many other civilizations existed prior to them, whether they were Terrestrial, ExtraTerrestrial or Inner Earth. We therefore osmotically absorb the gifts of all intelligences and fuse them into a Universal Mathematical Language. There is no Copyright allowable here. It belongs to the Public Galactic Domain.

The Ultimate Majik Square.

The **CONSTANT = 0** i.e: The Sums of the 3 columns, rows + 2 diagonals all equal 0.

(Jain 108's Ultimate Magic Square,
where the Constant or Magic Sum is **Zero**.
It uses the 9 essential numbers spanning from:

-4, -3, -2, -1, **0**, 1, 2, 3, 4

which makes sense,
embracing the negative and positive realm of God's Numbers
cleverly arranged so that the 3 columns, the 3 rows
and the 2 main diagonals all add up to nothing!)

EXERCISE 6

Quickly determine the multiplication of all the squares in the nineties, using Mental, One-Line Arithmetic in Base 100.

a) 98 x 98 =

b) 97 x 97 =

c) 96 x 96 =

d) 95 x 95 =

e) 94 x 94 =

f) 93 x 93 =

g) 92 x 92 =

h) 91 x 91 =

i) 88 x 88 =

The Sub-Sutra in use here is "**LESSEN BY THE DEFICIENCY**" which means this must be ticked on your page of sub-sutras on page 19.

ANSWERS 6

a) 98 x 98	= 98 − 2 / _ 2^2	= 96 / 04	= 9,604
b) 97 x 97	= 97 − 3 / _ 3^2	= 94 / 09	= 9,409
c) 96 x 96	= 96 − 4 / _ 4^2	= 92 / 16	= 9,216
d) 95 x 95	= 95 − 5 / _ 5^2	= 90 / 25	= 9,025
e) 94 x 94	= 94 − 6 / _ 6^2	= 88 / 36	= 8,836
f) 93 x 93	= 93 − 7 / _ 7^2	= 86 / 49	= 8,649
g) 92 x 92	= 92 − 8 / _ 8^2	= 84 / 64	= 8,464
h) 91 x 91	= 91 − 9 / _ 9^2	= 82 / 81	= 8,281
i) 88 x 88	= 88 − 12 / _ 12^2	= 76 / $_{1}44$	= 7,744

Sutra: BY THE EXCESS

OVER THE BASE of 100
(NIKHILAM SUTRA **means General Multiplication)**
(for EASY MULTIPLICATION of NUMBERS ABOVE or BELOW a BASE LIKE TEN, HUNDRED, THOUSAND, MILLION)

EXAMPLE: Multiply 103 x 104

The secret to obtaining this answer mentally, in one or two seconds, is to recognize that there is a visual pattern. The answer will be "something" on the Left Hand Side (LHS) followed by "2 digits on the Right Hand Side" (RHS).
Using my Inner Mental Screen, I visualize the following:

______________ / __ __

When you can forsee this Pattern, the Future appears in the Now, its like the numbers just fall into their allowed place, as naturally as rain falls down like in the Introduction to the Matrix Movie where we see the Digital Rain. Its like we are contacting another part of our Highest Potential State, the Answer is already there, we just need to Access it.

We think now of only the 2 Excesses, saying to ourself:
How much is 103 above the Base of 100, and how much is 104 above the Base of 100.

We mentally raise those two digits of "3" and "4" in our Inner Mental Screen. This process relates to another Sutra called YAVADANAM which means "to temporarily raise up, or hoist up, like a flag" aka "**ON THE FLAG**". It just means that we are learning how to hold digits in our Memory Bank, kind of putting them aside until we need them. Rapid Mental Calculation is about strengthening this ability, and it stimulates the 3rd Eye.

I now select one of the numbers, say "104" and add to it the excess of the other number, say the "3" of the "103" .This gives the LHS of the solution.

Then I multiply those 2 excesses held in my Memory Bank, which is "3" x "4".

Now the Solution will begin to look something like this:

= 104 + 3 / 3 x 4

= 107 / 12

= 10,712

I prefer that you learn to do this mentally, as shown above, it only takes one or two seconds to do it in your head. Then you can learn the written format shown below.

I have found in classes that when I teach the following written form first, the students find it difficult to learn the mental format. So if you are a teacher, please bear this in mind.

The following written method will reveal another hidden pattern, regarding LHS part of the answer, "107" how it can be obtained by two similar sums:

103 + 3

104 + 4

---

107 / 12 Notice the **Cross Addition** or Diagonal Relationship between these 4 numbers:

103 + 3

104 + 4

(103 + 4) is the same sum as (104 + 3). Both sums = 107.

This "Cross-Wise" phenomenon, observed by the Indian Yogis 2,000 years ago, is the **origin of the Multiplication symbol "X"**.

BASE 100
ANOTHER EXAMPLE, INVOLVING THE CARRY-OVER

EXAMPLE: 113 x 114

My Base is 100, so the predictable pattern to be visualized is:
"something, with 2 digits on the Right Hand Side":

________________ / __ __

= 114 + 13 / 13 x 14
= 127 / $_{1}$82 (Notice "the Baby One", which will get carried over to the LHS)
= 127 + 1 / 82
= 128 / 82
= 12,882

Here it is again in its written format, but remember to first master the above technique, then adopt this written method as a second choice.

(**100**)

113 + 13
114 + 14
____
127 / 182 But we are only allowed 2 digits on the RHS, since our Base 1**00** has 2 Zeroes in it, so we must slide the 1 of the **1**82 over to the LHS.

= 127 + 1 / 82
= 128 / 82
= 12,882

EXERCISE 7

Multiply the following numbers using Base 100:

a) **104 x 106** =

b) **105 X 107** =

c) **108 X 109** =

d) **112 x 112** =

e) **111 x 134** =

f) **112 x 117** =

g) **114 x 115** =

h) **102 x 104** =

i) **101 x 108** =

j) **101 x 109** =

ANSWERS 7

a) **104 x 106** = 106 + 4 / 6 x 4 = 110/24 = 11,024

(nb: As a good mental habit, when considering the 2 excesses: 4 and 6, it is easier to add the smaller excess to the other number, even though both "cross additions" have the same sum:
ie, both 104 + 6 is the same as 106 + 4)

b) **105 X 107** = 107 + 5 / 7 x 5 = 112/35 = 11,235
c) **108 X 109** = 109 + 8 / 9 x 8 = 117/72 = 11,772
d) **112 x 112** = 112 + 12 / 12 x 12 = 124/$_1$44 = 125/44 = 12,544
e) **111 x 134** = 134 + 11 / 34 x 11 = 145/3/3+4/4 = 145/$_3$74
= 148/74 = 14,874
f) **112 x 117** = 117 + 12 / 17 x 12 = 129 / 17+2/7x2
= 129 / 19/$_1$4 = 129/$_2$04 = 131/04 = 13,104
g) **114 x 115** = 115 + 14 / 15 x 14 = 129 / 15+4/5x4
= 129 / 19/$_2$0 = 129/$_2$10 = 131/10 = 13,110
h) **102 x 104** = 104 + 2 / 4 x 2 = 106/_8 = 10,608
i) **101 x 108** = 108 + 1 / 8 x 1 = 109/_8 = 10,908
j) **101 x 109** = 109 + 1 / 9 x 1 = 110/_9 = 11,009

DAY 7

- **Revision**
- **Cognate Numbers** eg: 22x28
- **Sutra: "By One More Than The Previous Digit"**
- **The Squaring of the Numbers Ending in 5** eg: 25^2
- Optional or Advanced Study:
  Sub-Sutra: **The First By The First and The Last By The Last**"
  eg: 26x86

SUTRA:

BY ONE MORE THAN THE PREVIOUS DIGIT

COGNATE NUMBERS

When Bharati Krsna Tirthaji wrote his book VEDIC MATHEMATICS he used words that most English speaking-people do not understand, having learnt a strict British system due to England's control of India at that time.
He skillfully introduced another concept of "**COGNATE NUMBERS**" to best describe the short-cut to
SQUARING NUMBERS THAT END In FIVE:
eg: 15 Squared = 1 x 2 / 5 x 5 =2/**25** based on the Sutra:
By One More Than The Penultimate Digit, (which used another big word that forced the English reader to pull out their Dictionary).
He wanted to explain why each time we squared a number ending in 5, like 25 squared = 6/**25**, and 35 squared = 12/**25** that the answer always ended in 25.

By Definition, "Cognate" means "**related to by Birth**".
What this means is that there is a special hidden pattern in the answer to the multiplication of double digit numbers, like

13 x	24 x	32 x	46 x	59 x
17	26	38	44	51

Can you see any patterning in those number pairs?

Look at these answers and determine for yourself the special short-cut that allows you to multiply all these numbers within seconds!
If you look closely you will observe that all the digits in the tens columns are the same. Notice that in the example above,

24 x
26 both the highlighted numbers are the same.

When also another condition is fulfilled, instant calculations can then happen, and that is when the both the numbers in the units column have a sum of 10, as in the same example

2**4** x
2**6**

both the highlighted numbers 4 and 6 have a sum of 10.

Thus 24 x 26 are Cognate Numbers in the sense that they are related to one another by their sameness of their digits in the Tens Column, and the sum of their digits in the Units Column both being 10.
When we have determined this factor we are able to perform a Rapid Mental One-Line Calculation.

Have a look at these four examples to see the overall pattern:

12	27	36	49	55	117
x	x	x	x	x	x
18	23	34	41	55	113
-----	-----	-----	-----	------	------

EXAMPLE COGNATE NUMBERS 24 x 26

24 x 26 = 2x3 / 4x6 = 6 / 24 = 624

EXAMPLE COGNATE NUMBERS 124 x 126

124 x 126 can be rewritten as:

12/4 x 12/6

This therefore illuminates the pattern or format for Cognate Numbers:

= 12 x 13 / 4 x 6

= 156 / 24

= 15,624

EXERCISE 1

Calculate, mentally, the following:

a) 12 x 18 =

b) 27 x 23 =

c) 36 x 34 =

d) 49 x 41 =

e) 55 x 55 =

f) 68 x 62 =

g) 74 x 76 =

h) 81 x 89 =

i) 97 x 93 =

j) 106 x 104 = clue = 10/6 x 10/4 =

k) 117 x 113 =

l) 125 x 125 =

ANSWERS 1

a)	12 x 18	= 1x2 / 2x8	= 2/16	= 216
b)	27 x 23	= 2x3 / 7x3	= 6/21	= 621
c)	36 x 34	= 3x4 / 6x4	= 12/24	= 1,224
d)	49 x 4	= 4x5 / 9x1	= 20/09	= 2,009
e)	55 x 55	= 5x6 / 5x5	= 30/25	= 3,025
f)	68 x 62	= 6x7 / 8x2	= 42/16	= 4,216
g)	74 x 76	= 7x8 / 4x6	= 56/24	= 5,624
h)	81 x 89	= 8x9 / 1x9	= 72/09	= 7,209
i)	97 x 93	= 9x10 / 7x3	= 90/21	= 9,021
j)	106 x 104	= 10x11 / 6x4	= 110/24	= 11,024
k)	117 x 113	= 11x12 / 7x3	= 132/21	= 13,221
l)	125 x 125	= 12x13 / 5x5	= 156/25	= 15,625

THE SQUARING OF NUMBERS ENDING IN 5

"Ekadikena Purvena" Sutra:
"By One More Than The Previous Digit"

The following Exercise is an extension of Cognate Numbers.
This is a lesson based on focusing on the Previous Digits or Penultimate Digits or those "in the shadow of" that led to the coining of the ancient revelation or Sutra now known as:
BY ONE MORE THAN THE PENULTIMATE DIGIT which simplified to:
BY ONE MORE THAN THE PREVIOUS DIGIT which contracted to the most widely used title of:
BY ONE MORE.
The Sanskrit title for this is called:
"Ekadikena Purvena".

EXAMPLE 15 x 15 or 15 Squared or 15^2

We are not concerned with the last digits, as we already know, from cognate number theory that all the answers will end in 25.
Our focus is only on the Previous Digit that is the 1 in **1**5 x **1**5 and we will multiply this by the next consecutive or sequential counting number which is 2, and tag on the 25, which gives the final answer:

15 x 15 = 1 **x 2** / 5 x 5 = **2**/25 = 225

EXERCISE 2

In the same format as the above example, solve these multiplications:

a) 25 x 25 =

b) 35 x 35 =

c) 45 x 45 =

d) 55 x 55 =

e) 65 x 65 =

f) 75 x 75 =

g) 85 x 85 =

h) 95 x 95 =

I) 105 x 105 = clue = 10/5 x 10/5 =

j) 115 x 115 =

k) 125 x 125 =

ANSWERS 2

a) 25 x 25 = 2x3 / 5x5 = 6/25 = 625
b) 35 x 35 = 3x4 / 5x5 = 12/25 = 1,225
c) 45 x 45 = 4x5 /5x5 = 20/25 = 2,025
d) 55 x 55 = 5x6 / 5x5 = 30/25 = 3,025
e) 65 x 65 = 6x7 / 5x5 = 42/25 = 4,225
f) 75 x 75 = 7x8 / 5x5 = 56/25 = 5,625
g) 85 x 85 = 8x9 / 5x5 = 72/25 = 7,225
h) 95 x 95 = 9x10 / 5x5 = 90/25 = 9,025
i) 105 x 105 = 10/5 x 10/5 = 10x11 / 5x5 = 110/25 = 11,025
j) 115 x 115 = 11/5 x 11/5 = 11x12 / 5x5 =
Use the Sutra: Digital Sums: When multiplying by 11 insert the sum of the two digits of the 12 between the two digits of the 12, thus in this case 11 x 12 = 1 / 1+2 / 2 = 1/3/2 = 132. Thus the final answer = 132/25 = 13,225
k) 125 x 125 = 12/5 x 12/5 = 12x13 / 5x5 = 12 + 3 / 2 x 3 / 25 = 15 / 6 / 25 = 15,625

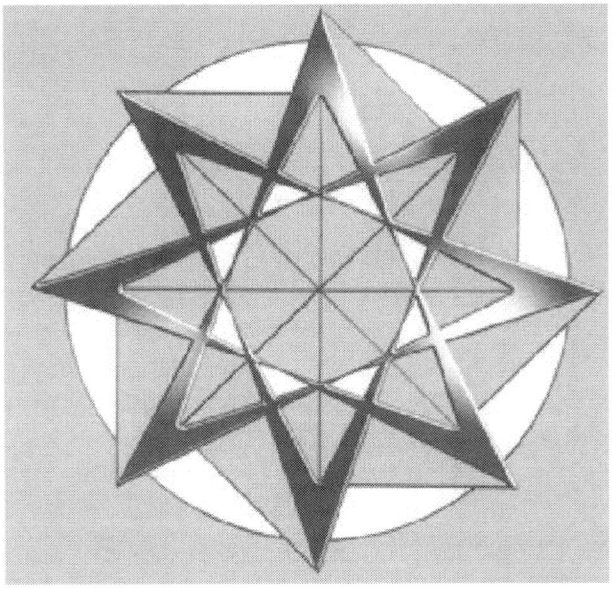

(**Jain 108**'s Logo for JAIN F.R.E.E.D.O.M.S
an acronym for:
For **R**esearch **E**xpressing **E**ssene **D**ata **O**f **M**agic **S**quares)

The FIRST BY the FIRST and the LAST BY the LAST

Sub-SUTRA: ADYAMADYENANTYAMANTYENA

Optional or Advanced Study

EXAMPLE 1: 26 x 86

This applies to the multiplication or product of any 2 digit numbers that obey 2 factors:
1 – the last figures or digits are the same and
2 – the first figures or digits add up to **10**.

26 x
86

Let us look at the example above: 26 x 86
The last digits or unit digits are both the same: 2**6** x 8**6** since both numbers end in 6. And the first digits are 2 and 8 which have a sum of **10**. Its kind of the opposite form to Cognate Numbers!

This means we can instantly and mentally calculate the answer by multiplying the first digit of each number (2 x 8) and adding the common last digit (6) which will give the first part or left hand side of the answer.

= (2 x 8) + 6 / __________

Then to get the last part of the answer, or the right hand side, just multiply the last digits which will always be a squared number: (6 x 6 or 6^2) = 36.

Thus the final answer = (2 x 8) + 6 / 36
= 22/36 = 2,236

[nb: This Sutra is almost the reverse of "By One More" when we multiplied COGNATE numbers like: 13 x 17 or 54 x 56].

EXAMPLE 2: 42 X 81

Using Two different Sutras:

PROPORTIONATELY and
The FIRST By the FIRST and the LAST BY the LAST.

42 x 81 = 2 x 21 x 81 = 2 x (21 x 81)
Thus the answer will be double (21x 81) = double (2x8)+1 / 01^2
= double 17 / 01
= 34 / 02
= 3,402

(**nb**: Go to page 29, your Sutra Page, and tick it to show you know it).

EXERCISE 3

a) 32 x 72 =

b) 25 x 85 =

c) 93 x 13 =

d) 64 x 44 =

e) 41 x 61 =

f) 86 x 26 =

then using the Sub-Sutra: **Proportionately**:

g) 62 x 71 =

h) 31 x 42 =

ANSWERS 3

a) 32 x 72	= (3x7)+2/2x2	= 23/_4	= 23/04	= 2,304
b) 25 x 85	= (2x8)+5/ 5x5	= 21/25	= 2,125	
c) 93 x 13	= (9x1)+3/3x3	= 12/_9	= 12/09	= 1,209
d) 64 x 44	= (6x4)+4/4x4	= 28/16	= 2,816	
e) 41 x 61	= (4x6)+1/1x1	= 25/_1	= 25/01	= 2,501
f) 86 x 26	= (8x2)+6/6x6	= 22/36	= 2,236	
g) 62 x 71	= 2 x (31x71)	= 2x[(3x7)+1/1x1] = 2x(22/01)		
	= 44/02			
h) 31 x 42	= 31x21x2	= ½(62x42)	= ½[(6x4)+2/2x2]	
	=½(26/04)	= 13/02	= 1,302	

DAY 8

- **Sutra**: **Transpose & Apply** eg: 34x5 reviewed as 34x10 then ÷ by 2
- Use other examples eg: 36x50 and 88x500
- **Compare** this to **Trachtenberg Method** eg: 34x5

The Sutra: Transpose and Apply, uses an opposite type of action to what is given. If we are told to multiply by 5, we can do a similar though opposite action by multiplying by 10 then dividing by 2.

EXAMPLE 34 x 5

We know that 34x5 is a 3Dx1D Trachtenberg type of calculation, but here we are challenged to do it another way, by Transpose and Apply, and determine which of the two styles or sutras serve to deliver the answer in the fastest mental way.

Using Transpose and Apply,

34 x 5 = (34 x 10) ÷ by 2
= 340 ÷ by 2
= 170

Using Trachtenberg Method 34x5 = 3x5/4x5 = 15/20 = 17/0 = 170
Both methods appear to take the same amount of time.

EXAMPLE 36 x 50 Using Transpose and Apply

36 x 50 = (36 x 100) ÷ by 2
= 3600 ÷ by 2
= 1800

Actually, the quicker way to do this is to immediately half the 36, which 18, and add on two zeroes.

Using Trachtenberg Method (36x5)x10 = 3x5/5x6 = (15/30)x10 = (18/0)x10 = 1800.
It appears that Transpose and Apply takes only a few seconds to multiply, a bit quicker than the Trachtenberg Method.

EXAMPLE 88 x 500 Using Transpose and Apply

Since 500 is half of 1000, I tag 3 zeroes onto half of 88 which is 44, giving the instant answer of 44,000. Done at the Speed of Thought.

EXAMPLE 88 x 25 Using Transpose and Apply

Since 25 is a quarter (1/4) of 100, I tag 2 zeroes onto a quarter of 88 which is 22, giving the instant answer of 22/00 or 2,200.

When solutions like this are done instantaneously, it relates to the highest or most supreme of all vedic sub-sutras called:
BY MERE INTUITION aka By Mere Observation aka By Mere Inspection. It Sanskrit, it is called: Vilokanam.

(On page 29, you can tick or mark this sub-sutra).

This means you just look at the Question, recognize the pattern via the Right Feminine Brain, and just call out the answer, no thinking, no logical mind, just pure Knowing or Intuition.
The aim of this booklet or tutorial is to get the student to this level of attainment, By Mere Intuition. This is the Gift of the Ancients. This will exercise the brain and stimulate all-remembrance of the Akashic Records, the Vibrational Library of all History and Herstory and Spirit Record of all Emotions ever emoted. It's another way of going into the twin spiral of our DNA, remembering who we are.

EXERCISE 1 USING TRANSPOSE & APPLY

a) 42 x 5 =

b) 63 x 5 =

c) 46 x 50 =

d) 78 x 50 =

e) 32 x 500 =

f) 75 x 500 =

g) 84 x 25 =

h) 22 x 250 =

i) 64 x 125 =

ANSWERS 1

a)	42 x 5	= (42x10)÷2	= 420÷2	= 210
b)	63 x 5	= (63x10)÷2	= 630÷2	= 315
c)	46 x 50	= (46÷2) x 100	= 2300	= 2,300
d)	78 x 50	= (78÷2) x 100	= 3900	= 3,900
e)	32 x 500	= (32÷2) x 1000	= 16000	= 16,000
f)	75 x 500	= (75÷2) x 1000	= 37.5 x 1000	= 37,500
g)	84 x 25	= (84÷4) x 100	= 21 x 100	= 2,100
h)	22 x 250	= (22÷4) x 1000	= 5.5 x 1000	= 5,500
i)	64 x 125	= (64÷8) x 1000	= 8 x 1000	= 8,000

DAY 9

Universal Sutra: "Vertically and Crosswise"

- Revision
- **2 D**igits x **2 D**igits: "Developing The Inner Mental Screen"
- First Example **12 x13** involving no Carry-Over
- Second Example **13 x14** involving a Carry-Over
- 2 other Examples: **35 x 35** and **23 x 81**
- **3 D**igits **x 3 D**igits eg: 123 x 456 and 144 x 035
- **Gelosia**: How did The Arabs use Cross-Grids to solve Teen Numbers, all 2 D x 2 D Numbers and 3 D x 2 Digit Nos. eg: 364 x 23

- Optional or Advanced Study
  to be studied in next booklet known as Level 2:
  - **3 D**igits **x 2 D**igits: eg: 144 x 35 using Moving Multiplier Method
  - **4 D**igits **x 4 D**igits: eg: 1234 x 4321
  - Worksheet on penciling in the actual "**X**" **shapes** up to 5 **D** x 5 **D**igits

Mural by Jain 1999 "Dolphusion".
Thanks to Aaron Lutze of the Gold Coast, for his computer skills designing this dvd cover.

Sutra:
VERTICALLY AND CROSSWISE

Notice how this array of Multiplications from ancient Bharat or India appears to resemble some Star Code or Language. This is how Multiplication was always done.
In fact, the very **origin of the Multiplication symbol**, when we write say 12x34

1 2 **x**
3 4

originates from this Sutra called Vertically and Crosswise. It is a memory of the diagonal relationships of those 4 digits, when we cross multiply (1x4) + (2x3).

This highly sophisticated form of Vedic Mathematics was practiced by the Aryans of the Upper Indus Basin some time before 1,000 BC. Some techniques were transmitted to Europe through the Arabs at the Court of Baghdad in 770 AD, to become the basis of much European mathematics. But more never got through, and it remained for the pope Shankar-archarya of Govardhana-Puri (ie: Swami Bharati Tirthaji 1884-1960) to rediscover the complete system of India through intensive study of certain Vedic texts.

The true honour of the invention of Zero, the decimal point and the place value system, and the concept of negative numbers all belong to Hindu roots, without which we would have no modern computers nor technology! (Other older civilizations like Babylonia etc also used zero, so this issue is very debatable).

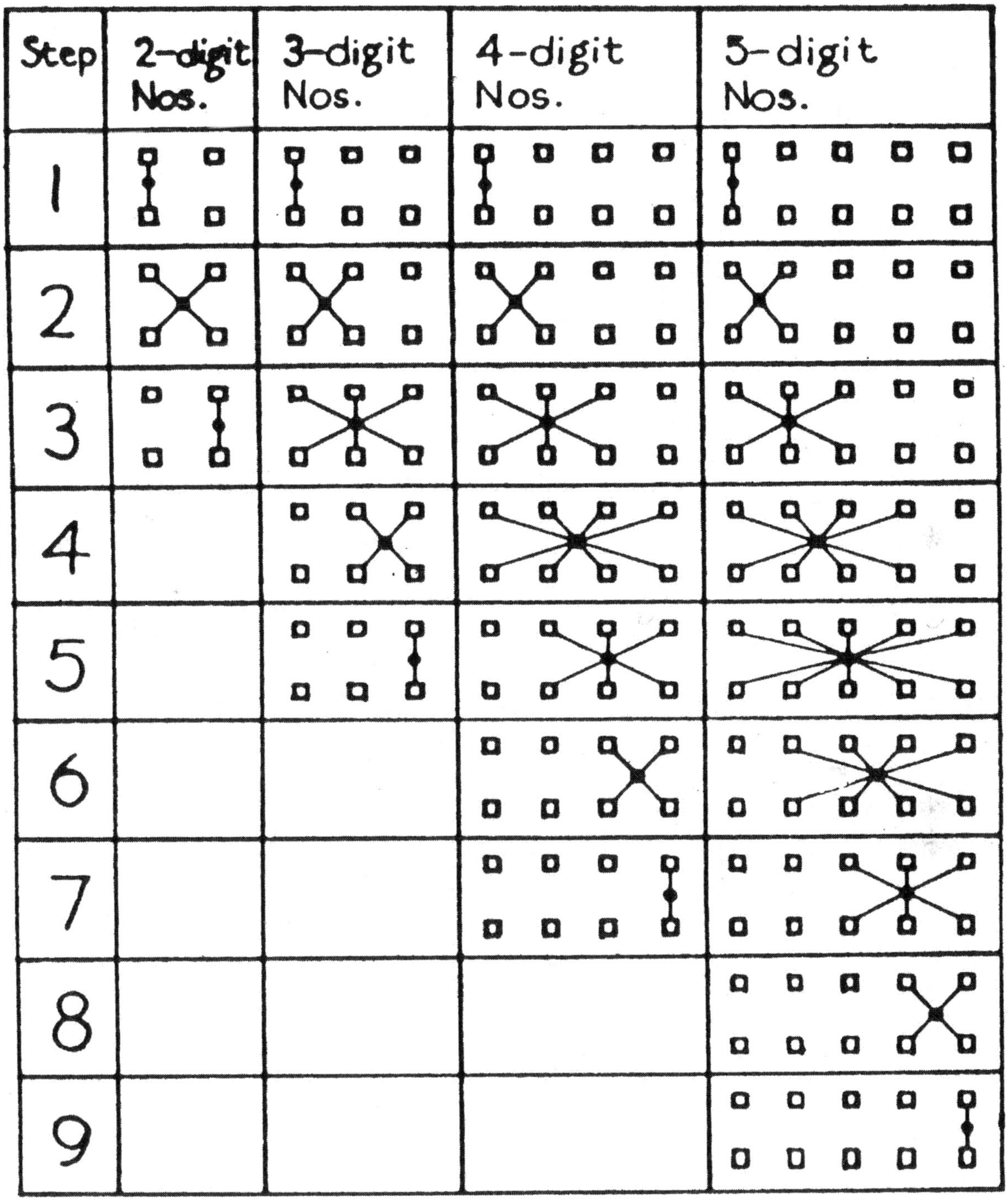

(Diagram is taken from: "Maths or Magic" by Joseph Howse who wrote a small booklet on Vedic Mathematics in the 1970s).

SUTRA: VERTICALLY AND CROSSWISE

aka: URDHVA-TIRYAK
Or Urdhva-Tiryagbhayam Sutra in Sanskrit.
(The UNIVERSAL SUTRA for all TYPES of MULTIPLICATION, ESPECIALLY WHEN the NUMBERS are NOT NEAR an EASY BASE)

EXAMPLE 12 X 13 Involving No Carry-Over

12 x
13

1x1 / (1x3) + (1x2) / 2x3

= 1 / 3 + 2 / 6
= 1/5/6 = 156

If you were to do this mentally, the mental working out would be like this:

"I multiply vertically on the right which is 2x3=6, then for the central section, I cross-multiply the two diagonals which gives 1x3 and 1x2 which gives 3 and 2, so I add them giving me 3+2=5, then for the extreme LHS digit, I multiply vertically on the left which gives 1x1=6. This gives me my Mental, One-Line Answer of 1/5/6".

nb: This Sutra is Ambidextrous in the sense that you can choose to calculate from the RHS, as in the case above, or do it from the LHS.
In India, most children do it from the LHS, but I find here in the west, that children prefer to do it from the RHS and working back to the LHS. It is your choice.

An Example Involving the Use of a Carry-Over:
Nb: This Sutra is generally performed mentally, so what follows is merely the written form of the mental form. You are only practicing the written form until you are able to master this Sutra via mental powers alone.

EXAMPLE 13 X 14 Example Involving A Carry-Over

```
  1 3 x
  1 4
------
1 7 2
 1
------
1 8 2
```

The mental working out would be:
"I multiply vertically on the RHS which is 3x4=12 or 3x4=$_1$2. I put down the "2" on the first row and carry over the "baby 1" in the second row. I make sure that I place this "baby 1" directly under the next future position which is the cross-multiplication of the 1x4 and 1x3=4+3=7. I need to add this "baby 1" to the 7 obtained, thus 7+1=8, I put down the "8" then multiply vertically on the LHS which is 1x1=1, forming my answer of 182.

EXAMPLE 35 X 35

```
    3 5 x
    3 5
   _____
  9 0 5
  3 2
 ______
 12 2 5
```

The beauty of this Sutra is that it is the Universal One, the one that is most generally applied, and solves all forms of multiplication. Remember in the previous chapter when we used the Nikhilam Sutra which is about multiplying numbers above or below an easy base, like 13 x 14, which are both near the Base of 10, we would use the Nikhilam Sutra and apply the Sutra: By The Excess.

But what happens when the two numbers being multiplied are not near an easy common base, eg: the numbers 23 x 81. We have arrived at a universal principle, a way of calculating by understanding this "X-Factor".

3   5
  X
3   5

The Cross-Multiplication of the above example 35x35 is the sum of 3x5 + 3x5 giving 15+15=30. When you consider this relationship of those 4 digits, how they are related to one another by these 2 diagonals, it forms a cross, or an "X". The deep inspiration here is that **have you ever wondered why you use this symbol "X" for multiplication**. Because it is derived from this original Sutra called Vertically and Crosswise.

This means that our ancient vedic seers, our forbears knew this intimate relationship between numbers and naturally adopted this symbol: "X".

(In later studies with Squaring of Numbers and its Inverse of Square Roots, we will be using these sum of the two diagonals which is called **DUPLEX** and written here as D(35) = 15+15 = 30).

This calculation means also that you have a choice of several sutras to use in some cases, like the previous example of 35 x 35 could be solved be using either 3 sutras:

1- **Vertically and Crosswise**, takes about 10 seconds to calculate mentally, or

2- **By The Excess** (Using Base 30 = 3x[35+5] / 5x5 = $120/_{2}5$ =1225), which takes about 8 seconds to multiply, or

3- **By One More** (3x4/5x5 = 12/25). This Sutra is the quickest and therefore the best choice of the 3 Sutras to use. It takes only 1 second to calculate this answer using this special formula. So your task is to get familiar with all the Sutras and quickly establish which formula is the best choice. This comes with practice.

EXAMPLE 23 X 81

23 x
81

= 2x8 / (2x1)+(3x8) / 3x1

= 16 / 26 / 3
= 18 / 6 / 3
= 1,863

"YOU NEVER CHANGE THINGS
BY FIGHTING THE EXISTING REALITY.
TO CHANGE SOMETHING,
BUILD A NEW MODEL
THAT MAKES THE EXISTING MODEL OBSOLETE."

- Buckminster Fuller

"THE KNOWLEDGE OF WHICH GEOMETRY AIMS IS THE KNOWLEDGE OF THE ETERNAL."
Plato

(Art by Jain 108)

EXERCISE 1

Multiply the following 2 Digit by 2 Digit numbers, using the sutra: Vertically and Crosswise:

a) 14 x 17 =

b) 21 x 34 =

c) 33 x 66 =

d) 42 x 65 =

e) 58 x 72 =

f) 64 x 91 =

g) 76 x 84 =

h) 85 x 43 =

ANSWERS TO EXERCISE 1

a) 14 x 17 =	1x1/(1x7)+(4x1)/4x7 = 1/11/28	= 1/13/8	= 2/3/8	= 238
b) 21 x 34 =	2x3/(2x4)+(1x3)/1x4 = 6/11/4	= 7/1/4	= 714	
c) 33 x 66 =	3x6/(3x6)+(3x6)/3x6 = 18/36/18	= 18/37/8	= 21/7/8	= 2,178
d) 42 x 65 =	4x6/(4x5)+(2x6)/2x5 = 24/32/10	= 24/33/0	= 27/3/0	= 2,730
e) 58 x 72 =	5x7/(5x2)+(8x7)/8x2 = 35/66/16	= 35/67/6	= 41/7/6	= 4,176
f) 64 x 91 =	6x9/(6x1)+(4x9)/4x1 = 54/42/4	= 58/2/4	= 5,824	
g) 76 x 84 =	7x8/(7x4)+(6x8)/6x4 = 56/76/24	=56/78/4	= 63/8/4	= 6,384
h) 85 x 43 =	8x4/(8x3)+(5x4)/5x3 = 32/44/15	= 32/45/5	= 36/5/5	= 3,655

DEVELOPING THE INNER MENTAL SCREEN

The **Cross-Over of the Optical Nerves** resembles closely the Sutra: **Vertically and Crosswise**. The Left Eye is connected to the Right Side of the Brain and the Right Eye is connected to the Left Side of the Brain. Thus from a bird's-eye view, the biology of our optic nerves is the shape of the "X" Sutra!

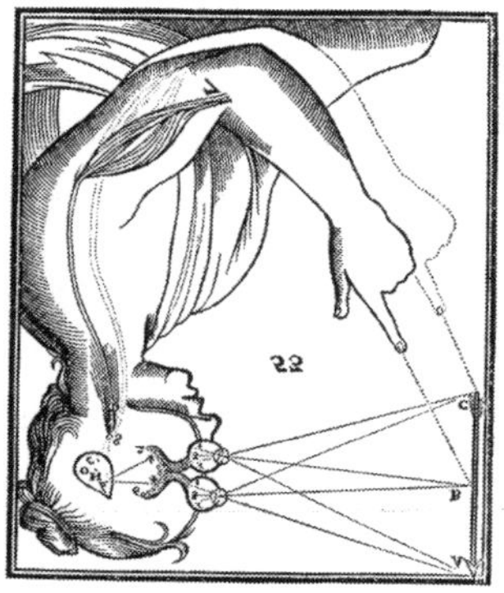

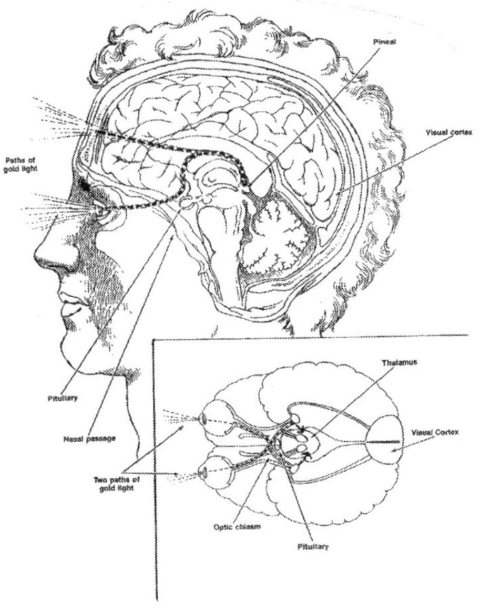

The Anatomy of Internal Viewing

(Above picture on right is reproduced from Barbara Brennan's "Hands Of Light" a medical Intuitive who could describe the inner organs of the human anatomy via clairvoyance).

Maharishi Mahesh Yogi states that **the 16 Vedic Sutras are Processes of the Mind** and invites students to meditate upon the deeper meaning of these Sutras.

eg: where else in Nature or Physics does this "X" shape appear. Make a list on the following page:

1 - Magnetic Fields: North and South Poles. **2 - Rosalind Franklin**, a woman who first discovered **DNA** by taking the **first radiograph** of the DNA molecule, exposed this "**X**" shape pattern.

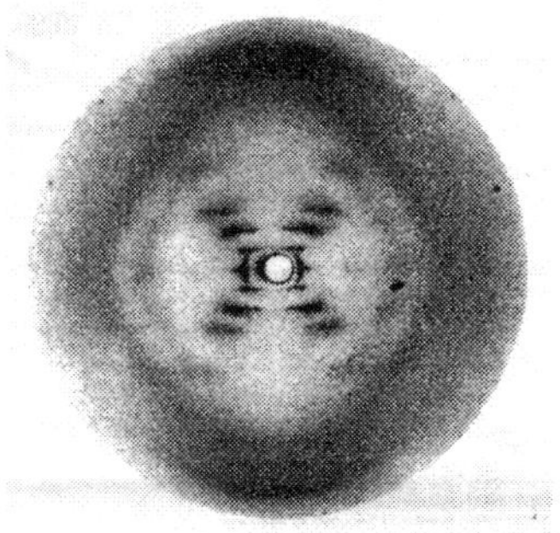

3 - (**Cross-section of the Tube Torus Donut showing nesting Tori** radiating in wavelengths of the Golden Ratio of 1:1.618033 therefore creating an infinite series bridging the Macrocosm to the Microcosm. Notice the "**X**"-shape or "X"-factor!).

EXAMPLES OF THE "X" SHAPE IN NATURE

"horny coral"

(displays the living form of the "X"-Shape: (taken from 'Art Forms In Nature' by Ernst Haeckel).

Maharishi Mahesh Yogi, says that the 16 Sutras and 14 Upsutras (Sub-Sutras) are really Processes of the Mind and warrant deep Meditation.
Regarding this Sutra: Vertically and Crosswise, can you make a list of examples where this "X" Factor appears in the Physical World or Living Nature, or Crystals/Minerals, or Physics, Science etc.

1. ....
2. ....
3. ....
4. ....
5. ....
6. ....

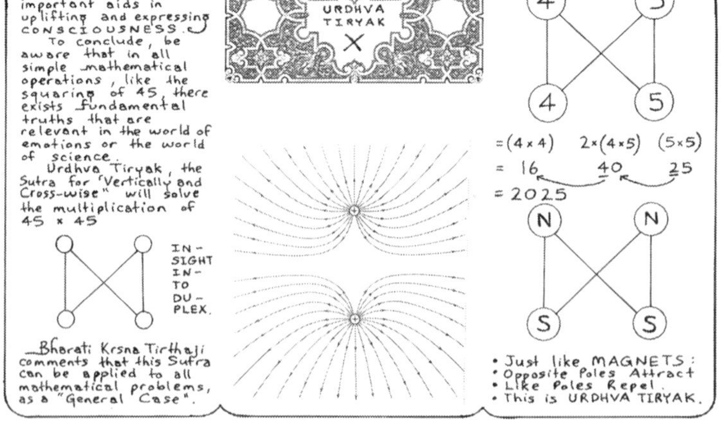

GELOSIA MULTIPLICATION

HOW DID THE ARABS PERFORM THEIR TEEN NUMBER MULTIPLICATIONS? eg 14 x 16

+

(Multiplication of 3 Digits By 2 Digit Numbers)

Aka: **LATTICE MULTIPLICATION**

This is how the Arabs were doing a clever though somewhat complex multiplication, at the same time when the Hindus were doing it effortlessly in their head! It is still an intelligent method, but I introduce it here as an example of MULTI-CULTURAL MATHEMATICS. Now we know 3 various and interesting systems:

1 – The British Maths System that works vertically down the page
2 – The Vedic System that calculates horizontally in the mental screen, and
3 – The Arab System using cross-grids.

(The origin of the word "**Gelosia**" or "**Jalousia**" is hazy, but some scholars say that it is the Italian word for "Jealousy", as the Shah's or noblemen of Arabia at that time guarded their many wives from the desirous eyes of others by having them safely behind iron or stone carved grills that appear like this latticework shown).

To understand how 14 x 16 was achieved a thousand years ago, using this cross-grid, study the pattern on the next page for 36.4 x 2.3 that gives an answer of 83.72. It is really using a Lattice Grid for multiplying 3 D x 2 Digit Numbers.

EXERCISE 2

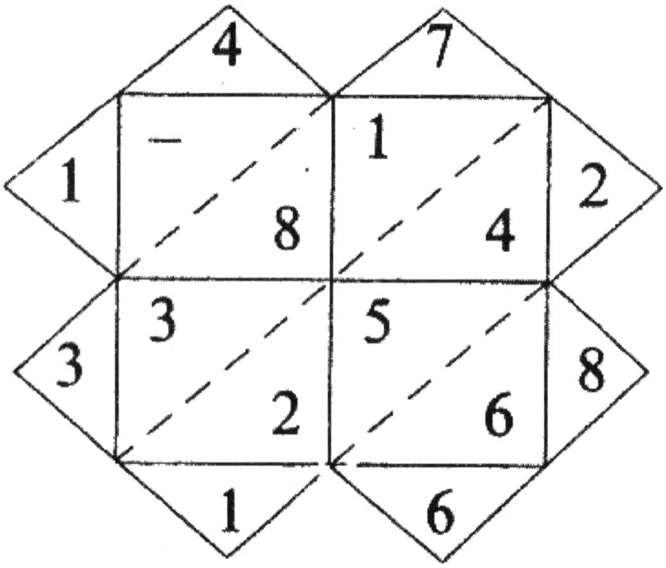

This diagram above shows the Arabic Lattice Multiplication of 47 x 28, which = 1,316.

Can you back-track and use this cryptic information to determine how to multiply your teen numbers like 14 x 16 ?

Here is the first part of the grid, for you to determine the answer.

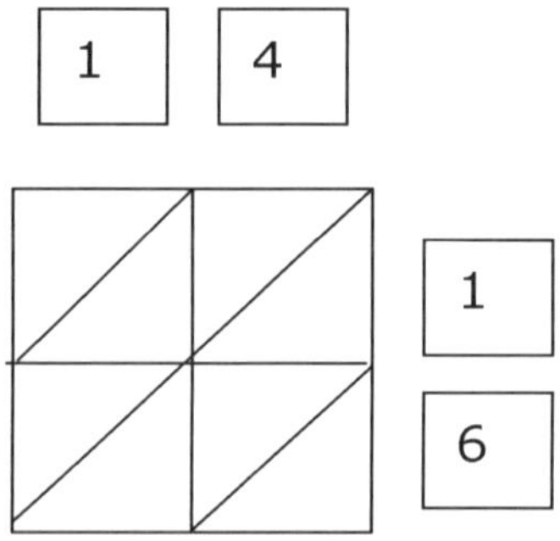

Answer to Exercise 2: 14x16=224

Gelosia multiplication

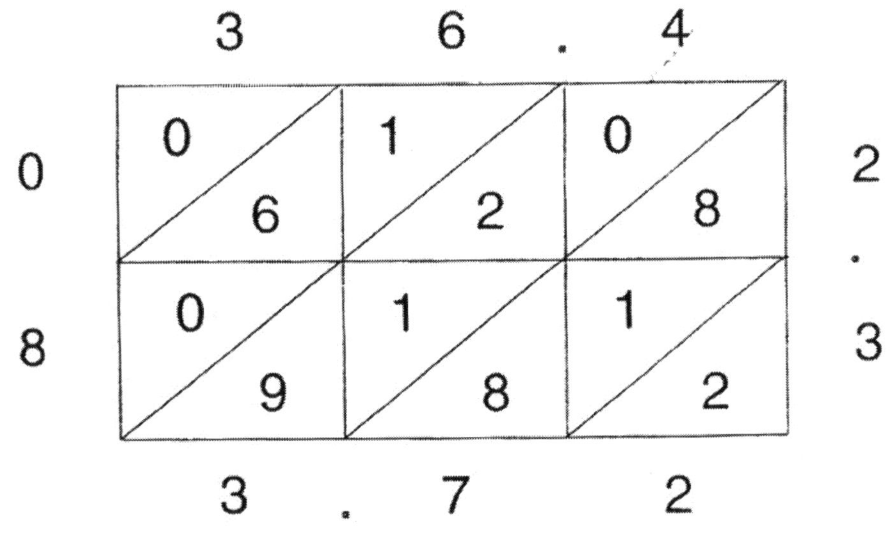

This technique was used in India from the 12th century onwards. The diagram shows the sum 2.3×36.4. The number 36.4 is written along the top edge, and 2.3 is written down the right edge. Each digit at the top is multiplied in turn by each digit down the side, and the answers are written in the boxes, with tens in the top left of each box and units in the bottom right corner. Next, the numbers along each diagonal are added up, beginning at the bottom right corner, with the units written below the diagonals and the tens carried over to the next diagonal. Two lines are drawn from the decimal points to a diagonal that points to where the decimal point should go. The answer is 83.72.

Vertically & Crosswise: 3 D X 3 D
MULTIPLYING 3-DIGIT BY 3-DIGIT NUMBERS

As explained before, this sutra means that we can solve calculations by applying a picture! This creates whole-brain learning, in the sense that numbers are logical Left-Cortex Brain, and a picture, shape or symbol is in the symbolic realm of the visual Right-Cortex Brain. Due to this superiority, it is predicted that the world's curriculum is shifting towards this method of multiplication, after the die-hard professors surrender to this fact. The billions of dollars involved in the re-writing of the world's mathematical curriculum is the main obstacle and most conventional teachers are happy getting paid teaching the old fossils, they have no real passion to bring in the new frequency. On the political or cultural scene, it may also be embarrassing for the elite Western Scholars to admit that the Indian Race have and have always had the best system.

In time though, it will be self-evident that this use of Cross-Multiplication Method of Vedic Mathematics was always the original method of multiplying large numbers.

Thus, what is the picture that will solve 3D x 3D? and how many glyphs or digits will the answer involve.

We let n = 3 as the question involves 3 Digit Numbers.
Double this number and less by 1, which is the same thing as saying: (2n-1). This equals 5, which means the solution will have a minimum of 5 digits in the answer.

The Pattern or Template or Master Arrangement will look like this:

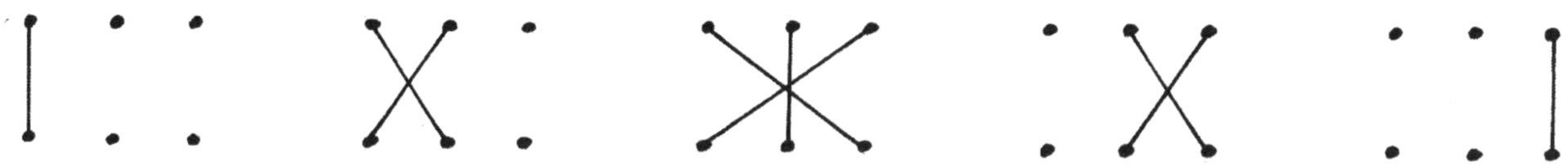

Step 5 **Step 4** **Step 3** **Step 2** **Step 1**

EXAMPLE 123 x 456

These are the 5 steps required to perform this as a mental, one-line calculation:

Step 1- Vertically on the Right: 3x6
Step 2- Two Cross-Mutliplications on the Right: (2x6)+(3x5)
Step 3- Three Cross-Mutliplications involving all digits:
(1x6)+ **(2x5)** +(3x4)
Step 4 - Two Cross-Mutliplications on the Left: (1x5)+(2x4)
Step 5 - Vertically on the Left: 1x4

1 2 3 x
4 5 6

1x4 / (1x5)+(2x4) / (1x6)+ **(2x5)** +(3x4) / (2x6)+(3x5) / 3x6

Notice that the axis of symmetry is at **(2x5)** shown in bold above and that there are 3 cross-multiplications to be added.
= 4 / 13 / 28 / 27 / 18
= 4 / $_{1}3$ / $_{2}8$ / $_{2}7$ / $_{1}8$
= 4 / $_{1}3$ / $_{3}0$ / 8 / 8
= 5 / 6 / 0 / 8 / 8

EXAMPLE 144 x 35 = 144 x 035

In this example, think of the "35" as "035" so it is now in the form of 3 Digits x 3 Digits.

1 4 4 x
0 3 5

1x0 / (1x3)+(4x0) / (1x5)+(4x0)+(4x3) / (4x5)+(4x3) / 4x5

= 0 / 3 / 17 / $_{3}2$ / $_{2}0$
= 0 / 3 / $_{2}0$ / 4 / 0
= 0 / 5 / 0 / 4 / 0
= 5,040

EXERCISE 3

Multiply the following 3D x 3D numbers as one-line answers:

a) 214 x 735 =

b) 324 x 682 =

c) 112 x 933 =

d) 555 x 555 =

e) 245 x 333 =

f) 707 x 213 =

g) 456 x 32 =

h) 618 x 33 =

ANSWERS TO EXERCISE 3

a) 214 x 735 = 2x7/(2x3)+(1x7)/(2x5)+(4x7)+(1x3)/(1x5)+(4x3)/4x5
= $14/{}_{1}3/{}_{4}1/{}_{1}7/{}_{2}0$ =15/7/2/9/0 = **157,290**

b) 324 x 682 = 3x6/(3x8)+(2x6)/(3x2)+(4x6)+(2x8)/(2x2)+(4x8)/4x2
= $18/36/{}_{4}6/{}_{3}6/8$ = $18/{}_{4}0/9/6/8$ = 22/0/9/6/8 = **220,968**

c) 112 x 933 = 1x9/(1x3)+(1x9)/(1x3)+(2x9)+(1x3)/(1x3)+(2x3)/2x3
= $9/{}_{1}2/{}_{2}4/9/6$ = 10/4/4/9/6 = **104,496**

d) 555 x 555 = 5x5/(5x5)+(5x5)/ (5x5)+(5x5)+(5x5)/(5x5)+(5x5)/5x5
= $25/{}_{5}0/{}_{7}5/{}_{5}0/{}_{2}5$ = 30/8/0/2/5 = **308,025**

e) 245 x 333 = 2x3/(2x3)+(4x3)/(2x3)+(5x3)+(4x3)/(4x3)+5x3)/5x3
= $6/18/{}_{3}3/{}_{2}7/{}_{1}5$ = $6/{}_{2}1/5/8/5$ = 8/1/5/8/5 = **81,585**

f) 707 x 213 = 7x2/(7x1)+(0x2)/(7x3)+(0x1)+(7x2)/(0x3)+(7x1)/7x3
= $14/7/{}_{3}5/7/{}_{2}1$ = $14/{}_{1}0/5/9/1$ = 15/0/5/9/1 = **150,591**

g) 456 x 032 = 4x0/(4x3)+(5x0)/(4x2)+(6x0)+(5x3)/(5x2)+(6x3)/6x2
= $0/12/{}_{2}3/{}_{2}8/{}_{1}2$ = 0/14/5/9/2 = **14,592**

h) 618 x 033 = 6x0/(6x3)+(1x0)/(6x3)+(8x0)+(1x3)/(1x3)+(8x3)/8x3
= $0/18/{}_{2}1/{}_{2}7/{}_{2}4$ = 0/20/3/9/4 = **20,394**

DAY 10

- Main and Final **Revision Test called "MATHLETICS"**
- **End of Course**
- **Offering of "Certificate Of Completion"**
- **Video Testimonials by Parents**
- **Feedback Sheet by Students**
- **Books and Dvds by Jain** on Rapid Mental Calculation

END Of RAPID MENTAL CALCULATION LEVEL ONE
(JAIN MATHEMAGICS ONLINE CURRICULUM For The GLOBAL SCHOOL)
For TEENAGERS From AGES 13 To 19

RULES or GUIDELINES for the FINAL MATHLETICS TEST:

This final exam or test is not meant to be about competitiveness, it is offered as a mean to create classroom fun, and putting the students on focus in regard to remembering all their sutras and how to arrive with mental one-line answers.

- Before calling out the answer, first identify the name of the sutra then give the answer.
- One point is given for the Naming of the Sutra,
- One point is given for the Correct Answer.
- Select a person in the room as ScoreKeeper.
- In the classroom situation, once you have called out an answer, and it is not correct, you are not allowed to call out a second time; this gives the other students a chance to call out the answer. This also invites you to carefully think of your answer before calling it out.
- The questions will be selected in chronological order from the beginning of the book till the end, systematically going through Day 1 to Day 2 etc to Day 9.
- The prize for the student with the highest score will receive 2 dvds valued at $90 by Jain Mathemagics.... One called "Vedic Mathematics For The New Millennium", (the first complete dvd in the world on this topic) and an Introduction to Ancient Knowledge called "The Living Mathematics Of Nature", 1 of 5 of a 5 dvd set.

nb: This Test can be done as a written test so students are not interrupted, or as an Oral Test in Public with Parents viewing...

Revision Test: "MATHLETICS"

No.	NUMBER QUESTION	RELEVANT SUTRA	ANSWER
			DAY 1
1	9 x 6 =		
2	9 x 38 =		
3	8 x 7 =		
4	Recite fluently without stutter, your 6 x Table up to 6x12=72		
5	Recite fluently without stutter, the list of Sequence of Squared Numbers ie: 1^2, 2^2, 3^2 etc up to 9^2.		
6	11 x 33 =		
7	11 x 84 =		
8	11 x 745 =		
			DAY 2
9	1000 - 333 =		
10	How much change from $10 when spending $2.95?		
11	10,000 - 3,455 =		
12	1,000,000 - 285,496 =		
			DAY 3
13	48 + 48 =		
14	149 + 47 =		
15	219 + 219 =		
16	What sutra did the Indians use 2,000 years ago when doing their Times Table? Using this Sutra, solve 7 x 8 showing all the working out.		
17	Similarly, show all the working out for 7 x 6 =		
			DAY 4
18	36 x 6 =		
19	44 x 8 =		
20	617 x 4 =		
21	713 x 8 =		
22	What is 6 Cubed aka 6^3?		

23	Recite fluently without stutter, the list of Sequence of Cubed or Cubic Numbers ie: 1^3, 2^3, 3^3 etc up to 9^3.		
24	12 x 14 =		**DAY 5**
25	13 x 17 =		
26	16 x 16 =		
27	Showing the working out for 19x19 using Base 20		
28	22^2 =		**DAY 6**
29	21^2 =		
30	54^2 =		
31	58^2 =		
32	98 x 97 =		
33	98^2 =		
34	103 x 104 =		
35	111 x 125 =		
36	1012 x 1013 =		
37	22 x 28 =		**DAY 7**
38	71 x 79 =		
39	35^2 =		
40	95^2 =		
41	33 x 73 =		
42	34 x 5 =		**DAY 8**
43	36 x 50 =		
44	68 x 500 =		
45	64 x 25 =		
46	88 x 250 =		
47	64 x 125 =		
48	21 x 82 =		**DAY 9**
49	Multiply 21 x 82 using the standard British format.		
50	Multiply 21 x 82 using the Arabic Gelosia Method		
51	123 x 456 =		
52	618 x 33 =		

CONCLUSION:

Thus ends the first part of this course, Level One, taking us closer to a deeper matheological understanding of how the Universe works. (Matheology = ma**THEO**logy. "Theo" is an ancient Greek work meaning "God", so whatever this word "Matheology" means to you, I will let you decide).

Mathematics is therefore a Star Language, giving us the sense of "X-Ray Eyes" allowing us to see the inherent order in all the apparent randomness or chaos.
In this way it indicates that there is FIXED DESIGN, ETERNAL DESIGN, TIMELESS DESIGN. It's a Universal Language of Shape and Sacred Symbol that can be perceived or read by any entity on any planet and in any time, 100 years from now or 10,000 years from now; the script can not change as it is the Language, Blueprint or Template of Creation.

Revision Test: "MATHLETICS" with ANSWERS			
No.	**NUMBER QUESTION**	**RELEVANT SUTRA**	**ANSWER**
1	9 x 6 =	Looking at back of Hands, Finger Bend	**DAY 1** 54
2	9 x 38 =	Looking at back of Hands, Insert a V + Finger Bend	342
3	8 x 7 =	Magic Fingers: Looking at Palm of Hands	56
4	Recite fluently without stutter, your 6 x Table up to 6x12=72		6, 12, 18, 24, 30, 36, 42, 48, 54, 60, 66, 72.
5	Recite fluently without stutter, the list of Sequence of Squared Numbers ie: 1^2, 2^2, 3^2 etc up to 9^2.		1, 4, 9, 16, 25, 36, 49, 64, 81.
6	11 x 33 =	Digit Sums aka Digital Compression	363
7	11 x 84 =	ibid	924
8	11 x 745 =	"Add The Neighbour On The Right"	8,195
9	1000 – 333 =	"All From 9 & The Last From 10"	**DAY 2** 667
10	How much change from $10 when spending $2.95?	ibid	$7.05
11	10,000 – 3,455 =	ibid	6,545
12	1,000,000 – 285,496 =	ibid	714,504
13	48 + 48 =	"By The Completion"	**DAY 3** 96
14	149 + 47 =	ibid	196
15	219 + 219 =	ibid	438
16	What sutra did the Indians use 2,000 years ago when doing their Times Table? Using this Sutra, solve 7 x 8 showing all the working out.	= Sutra: "By The Deficiency"	56
17	Similarly, show all the working out for 7 x 6 =	ibid	42
18	36 x 6 =	Trachtenberg Method of 2D x 1D	**DAY 4** 216
19	44 x 8 =	ibid	352

20	617 x 4 =	Trachtenberg Method of 3D x 1D	2,868
21	713 x 8 =	Trachtenberg Method of 3D x 1D	5,704
22	What is 6 Cubed aka 6^3?	Ibid **6^3** = 6x6x6 = 36x6 = 18 / $_3$6 = 21/6 =	216
23	Recite fluently without stutter, the list of Sequence of Cubed or Cubic Numbers ie: 1^3, 2^3, 3^3 etc up to 9^3.		1, 8, 27, 64, 125, 216, 343, 512, 729
			DAY 5
24	12 x 14 =	"By The Excess" using Base 10	168
25	13 x 17 =	ibid	221
26	16 x 16 =	ibid	256
27	Showing the working out for 19x19 using Base 20	"By The Deficiency"	361
			DAY 6
28	22^2 =	"By The Excess" using Base 2	484
29	21^2 =	ibid	441
30	54^2 =		2,916
31	58^2 =		3,364
32	98 x 97 =	"By The Deficiency" using Base 100	9,506
33	98^2 =	"By The Excess" using Base 50	9,604
34	103 x 104 =	ibid	10,712
35	111 x 125 =	ibid	13,875
36	1012 x 1013 =	"By The Excess" using Base 1000	1,025,256
			DAY 7
37	22 x 28 =	Cognate Numbers. "By One More"	616
38	71 x 79 =	ibid	569
39	35^2 =	Squaring Of Nos. Ending In 5. "By One More"	1,225
40	95^2 =	ibid	9,025
41	33 x 73 =	"The First By The First and The Last By The Last"	2,409
			DAY 8
42	34 x 5 =	Sutra: "Transpose & Apply" or Trachtenberg	170
43	36 x 50 =	ibid (nb: 50 = 100/2)	1,800
44	68 x 500 =	Ibid (nb: 500 = 1000/2)	34,000 cont...

45	64 x 25 =	"Transpose & Apply" only. (nb: 25=100/4)	1,600
46	88 x 250 =	ibid	22,000
47	64 x 125 =	Ibid (nb: 125 = 1000/8)	8,000
48	21 x 82 =	"Vertically & Cross-Wise"	**DAY 9** 1,722
49	Multiply 21 x 82 using the standard British format.	Multiplication Going Down The Page	1,722
50	Multiply 21 x 82 using the Arabic Gelosia Method	Lattice Multiplication	1,722
51	123 x 456 =	"Vertically & Cross-Wise" 3dx3d	56,088
52	618 x 33 =	"Vertically & Cross-Wise" 3dx3d Visualize as 618x033	20,394

APPENDIX

(Art Of Jain, 1980,
(My first mural, 1mx1.5m
displaying a sense of innate geometry)

FEEDBACK SHEET

Name
and Age:

....

What is your
Profession?

....

Date:..

Name of Seminar(s):
- Vedic Mathematics aka Rapid Mental Calculation
- Magic Squares
- Divine Phi Proportion
- 3-Dimensional Geometry (5 Platonic Solids)
- Art Of Number

How would you rate this Seminar? (Please Circle: 1 is low, 10 is excellent)

1 2 3 4 5 6 7 8 9 10

What did you enjoy most about this Seminar?
Were there any special moments or peak experiences?

What criticism or improvement could you recommend to Jain to help in upgrading his next presentation?

Do you give permission that Jain can use any of the above short feedbacks as a TESTIMONIAL that can posted on Jain's website?

Yes No

Is there another person who you would like Jain to contact to recommend this course to. I will need their:
Name:
Phone Number:
Nearest Main City of Residence:
Email:

END OF COURSE

Issuing of laminated **CERTIFICATES** to STUDENTS.

Celebration

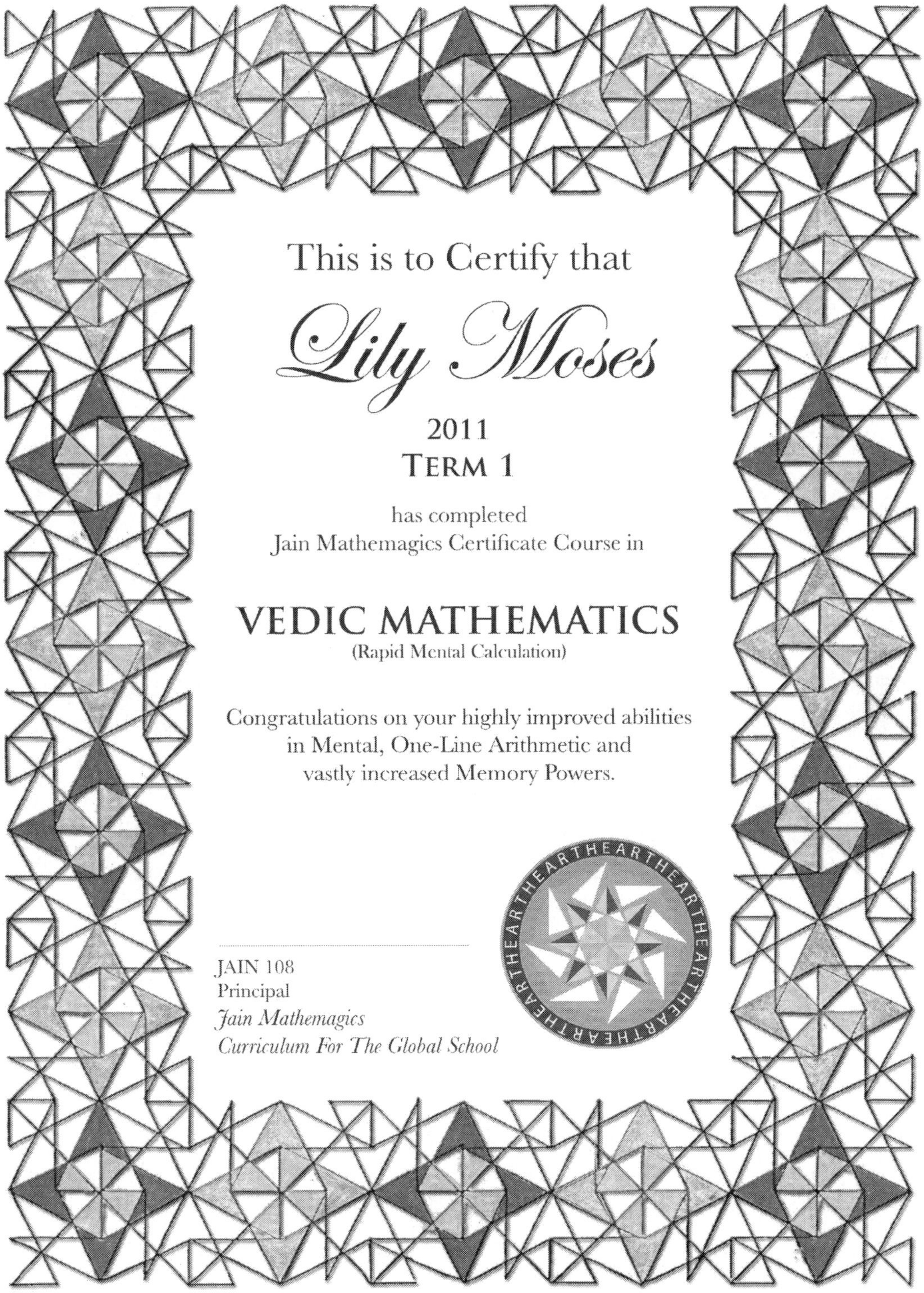

This is to Certify that

Lily Moses

2011

TERM 1

has completed

Jain Mathemagics Certificate Course in

VEDIC MATHEMATICS

(Rapid Mental Calculation)

Congratulations on your highly improved abilities
in Mental, One-Line Arithmetic and
vastly increased Memory Powers.

JAIN 108
Principal
Jain Mathemagics
Curriculum For The Global School

MATHEMAGICS FOR TEENS AND JUNIORS
TERM 3 of 2006

A REPORT by JAIN

The PARENT-CHILD RELATIONSHIP in VEDIC MATHEMATICS CLASSES

An interesting phenomenon has occurred for Term 3 of 2006 (that runs for 8 weeks, one class per week), regarding the separate or individual classes for children and adults. To increase afternoon class sizes at Shearwater Steiner School, I joined the 5 teenagers with the 5 adults. It was interesting that all the adults that had registered for their class were also the parents of the teenagers that had enrolled, so when the class combined, each Parent sat near their teenage child.
The same phenomenon happened with the Mathemagics for Juniors, it was expected that Parents would sit in as guests to be with their 8 to 12 year old child, so this was not surprising. In this Junior's class, 4 Parents sat with their child.

I can report that in both classes, but mainly with the teenage students, it became quite evident that the teenagers invariably calculated their multiplications at the speed of light, much faster than their parent who was still scratching their head or still attempting to work out the answer to common questions like 13x14 or 103x104.
It was quite satisfying to see this in play, as it demonstrates how the younger mind, not so indoctrinated by western style mathematics, could be more flexible enough to adopt a new mathematical method like Mental Calculation, whereas the Parents had been so locked into their old and more comfortable way of multiplying digits.

I also observed that the Parents were doing this class not just for the high benefit of their child, but because they are interested in working through their deep mathematical hurts or wounds or hang-ups that originated from their early years at school when they were told that they were not good enough, or dumb, or failures and a whole tirade of such abuse. The good news is that after only about 4 lots of 2 hour sessions, which is half of the Certified Certificate Course, they are realizing, thanks to the efficiency of the Vedic and Jain Mathemagics Methods, that they are brilliant mathematicians getting results at the Speed of Thought, thanks to their ability to tune into the supreme and lost Art of Pattern Recognition.

I am writing this report on Day 4 of the Day 8 (1 day per week) program as all Parents and Students have vastly improved their mathematical capabilities.
Even if a student does not choose to run with this Vedic Mathematics, it is still implanted within their Consciousness that there is another system at play, and at a later date, they may remember this information or make a reference to it that once they had learnt another way.

Obviously the challenge here is that the schools the students go to have not yet embraced that there is another way of multiplying, adding or dividing their digits. So this transition for the Global Curriculum to become embraced, may take a generation or so until the time comes when everyone naturally performs their mathematical calculations in their head.

I can only report or conclude that I was delightfully surprised to witness this evolving Parent/Child relationship and plan to promote the next upcoming Term 4 of 2006 Mathemagics classes in this vein of Dual Learning, inviting the Parent to become more involved in their Child's Mathematical Education.

Another surprise was to hear the 8 year old students demanding that the 2 hour class not end. Isn't that surprising, the child has been at school all day, then they do anther two hours of Mental Calculation with me and at the end of the class create a big fuss to keep going on, at which point the parent has had enough, brain-drained, and the lesson closes. I perceive that they see Mental Calculation as a game or a play, that is why I suspect that they like it so much.

As an aside, I did some research, asking of the Parents in my classes, whether or not their own Parents had had an active involvement in their growing years at school. Their answers were all a startling and common NONE or NIL.

Both Parents and Teenagers said that they enjoy quizzing or challenging one another when they go on a car, train or bus ride: "OK, Mum, you've got only 5 seconds to tell me what is 108 x 108?"

I will update upon this report after the students and parents have received their Certificates for completing the course. I believe that there is more to report or elaborate on, as other half-hidden emotions are surfacing or becoming apparent, like the deep bond and emotion of "proudness" of the Parent/Guardian toward the Child or vice-versa, due to this "caring", I sense riveting life-long connections forming; I sense "Gratitude" from both parties for experiencing this new mathematics in this non-threatening environment of "togetherness". There is also this vague sense of "sporting" as subtle challenges are manifesting between the Parent and Child to see who can out-wit one another.

Especially because this ancient knowledge is not well-known, it makes the knowledge being taught more sacred, and thus the Parent and Child are actually discovering a treasure in this state of being together, in this beautiful state of "sharing the experience". As you can see, this phenomenon has many layers to it, and needs many more months and perhaps years to reveal the depth of what has just been discussed.

With all this inspiration, I expect that my next promotional flyer would like something like this:

JAIN MATHEMAGICS and VEDIC MATHMATICS
DUAL or COMBINED LEARNING CLASSES
FOR TEENS and PARENTS (ages 13 and above)
FOR JUNIORS (ages 9 to 12) and PARENTS: TERM 4 of 2006

- Get involved with your Child's education
- Learn the supreme Art of Mental Calculation
- Develops your Mathematical Confidence
- Increases your Memory Power
- Dr Jain heals past Mathematical Traumas.

(for more information about the classes, click on: **http://www.jainmathemagics.com/page/3/default.asp**)

MATHEMAGICS 2007

The

Distributed by Jain Mathemagics

Living Mathematics Of Nature

5 DISC DVD SERIES

Introduction to Jain Mathemagics
An introduction to the 4 topics below.
Explains each topic, providing good overview of work.

Vedic Mathematics
Rapid Mental Calculation.
No more calculators. Increases your memory power!

Magic Squares
Translating numbers into Atomic Art.
Teaches children pattern recognition.

The Divine Phi Proportion
Explores the Geometry of Flowers
identical to the Human Canon!

3-Dimensional Geometry
The 5 Platonic Solids
adored by Pythagoras and his community.

Distributed by JAIN MATHEMAGICS
777 Left Bank Road Mullumbimby
NSW 2482 Australia

Ph: (02) 6684 4409
Email: jain@jainmathemagics.com
www.jainmathemagics.com

Videos & Books Available Online...

www.jainmathemagics.com

Jain Products Order Form

NEW! **The Living Mathematics of Nature, Series – 5 DVD Set, pub: 2006**		
☐ **Complete 5 DVD Set** (Special! Save $25)	$200	240 gm
☐ **Introduction to Ancient Knowledge** (DVD #1 of 5)	$ 45	100 gm
☐ **Vedic Mathematics** (DVD #2 of 5)	$ 45	100 gm
☐ **Magic Squares** (DVD #3 of 5)	$ 45	100 gm
☐ **The Divine Phi Proportion** (DVD #4 of 5)	$ 45	100 gm
☐ **3-Dimensional Geometry** (DVD #5 0f 5)	$ 45	100 gm
MAGIC SQUARES		
☐ **The Book of Magic Squares, Volume 1** (M. Sqs. of 3, 4, 5, 6, 7)	$ 50	520 gm
☐ **The Book of Magic Squares, Volume 2** (M. Sqs. of 8, 9, 10)	$ 50	520 gm
☐ **The Book of Magic Squares, Volume 3** (M. Sqs. of 11, 12, 16)	$ 50	560 gm
☐ **Join-The-Dots** (adults and children coloring book. Pub:1999, 56 pp)	$ 35	160 gm
☐ **Decal: The Magic Square of 4 X 4 (Jupiter)** (Small, Magenta)	$ 10	20 gm
☐ **Decal: Magic Square 8 X 8 (Mercury)** (Small, Blue Background)	$ 10	20 gm
☐ **Decal: Magic Square 4 X 4 (Jupiter)** (Large, Yellow Background)	$ 25	100 gm
☐ **The Gospel of the Holy Twelve** (Essene Bible w. Magic Sqs. 210pp)	$ 45	410 gm
VEDIC MATHEMATICS		
☐ **The Magic of Nine in Vedic Mathematics.** (Pub:1999, 124pp)	$ 50	390 gm
☐ **The Vedic Mathematics Curriculum for the Global School. Part 1, Digital Sums.** (Pub:2001, 142 pp)	$ 50	460 gm
☐ **Jain Mathemagics Curriculum for the Global School. Part 2. Multiplication.** (pub: 2005, 152pp)	$ 50	480 gm
☐ **Vedic Mathematics for the New Millennium: DVD** (Pub:2002)	$ 50	100 gm
☐ **Dolphin Human Connections**: Jain shows Vedic Maths (2006). 40 mins	$ 55	100 gm
☐ **Vedic Mathematics** by: Bharati Krsna Tirthaji **Soft Cov.** (355pp)	$ 45	460 gm
☐ **Vedic Mathematics** by: ibid. **Hard Cover.** (Pub:1965, 355pp)	$ 55	570 gm
☐ **Calculations @ The Speed Of Light** by Kranti Kiran (pub:2002. 104pp)	$ 45	230 gm
DIVINE PHI PROPORTION		
☐ **The Book of Phi, Vol 1** (The Living Maths of Nature. Pub:2002, 172pp)	$ 60	550 gm
☐ **In The Next Dimension** (The Book of Phi, Vol 2. Pub:2003, 188pp)	$ 60	580 gm
The Book of Phi, Vol 3 (Intro to the 108 Phi Code Pub:2009 pp216)	$ 65	550 gm
The Book of Phi, Vol 4 (The 108 Codes. Pub:2010, 172pp)	$ 65	550 gm
☐ **Propositions on the Exact Value of Pi.** Reddivari Sarva J. Reddy	$ 70	620 gm
☐ **Peter's Projection Map.** (Earth's true cartographic proportions)	$ 25	50 gm
☐ **Platonic Solids:** Crystal Gift Set in wooden camphor boxes	$100	50 gm

EBOOKS available of all books

SUB-TOTAL: $ ....

Postage & Handling: $ ....

TOTAL: $ ....

Tax Invoice

Name: ....

Mailing address: ....

City/Town: .... **State**: .... **Zip Code**: ....

E-mail Address: ....

Home phone: ....**Mobile**: ....

Please choose your preferred method of payment:

☐ **Cash, Cheque or Money Order.** Amount: ....
Payable: JAIN F.R.E.E.D.O.M.S.

☐ **PayPal now Available!** Amount: ....
Email: jain@jainmathemagics.com

☐ **Direct Deposit** Amount: ....
Acc Name: JAIN F.R.E.E.D.O.M.S. BSB: 032573 Acc Number: 171268

☐ **(circle) Visa / Mastercard** Amount: ....

Card number..

Name on Card: .... **Exp. date**: ....

AUSTRALIAN POSTAGE: Orders up to $50, add $8. Orders $50 to $120, add $12.

INTERNATIONAL ORDERS: $10 set up fee and add $10 per item. (Approx.)
International Bank Draft in Australian Dollars. PayPal also available + Western Union.
Cheques made out to JAIN F.R.E.E.D.O.M.S.

All Prices are listed in Australian Dollars. **All products available online!**

Date Of Order .... **Dispatched** ....

6	1	8
7	5	3
2	4	9

JAIN F.R.E.E.D.O.M.S. inc 9882763 trading as JAIN MATHEMAGICS
For Research Expressing Essential Data Of Magic Squares
Phone: +61 02 6684 4409 Mobile: +61 0423 583 886 ABN 47 244 872 696
Email: jain@jainmathemagics.com Web: www.jainmathemagics.com
Address: Pob 729, Mullumbimby, NSW, 2482, Australia

VEDIC MATHEMATICS SACRED GEOMETRY

Jain MatheMagics

MAGIC SQUARES ANCIENT KNOWLEDGE

Jain's Interactive Workshops will help you discover

Maths as Art

Maths as Science

Maths as History

Find out what you didn't learn in school...

Vedic Mathematics (Rapid Mental Calculation)

Find out how to accomplish complex and rapid calculations...in your head!

Used by NASA but not released to the public, Vedic Mathematics is on the leading eday of rediscovered technology - an ancient yet simple system of mental one-line arithmetic.

Jain will show you the 16 basic Sutras - Word Formulae - which can solve all known mathematical problems.

You will learn to multiply 2-digit x 2 Digit numbers, like 24 x 24 in your head... with no calculator, pen or paper.

Magic Squares (Translating Numbers into Atomic Art)

The Ancients discovered internal symmetries and harmonies that lay hidden within the chaos...elegant and PHi-Ratioed patterns,... atomic structures often symbolic of life's journey.

The Magic Square of 3x3, known as the Lo-Shu, in ancient China, is the centre of the Sino/Tibetan Cosmology. The sums of the columns, rows and diagonals all add up to 15, symbolically creating Order amidst the Chaos, and Equality in all dimensions.

Jain will show you how to create complex and intricate atomic art from numbers in various Magic Squares.

Three-Dimensional Sacred Geometry (5 Platonic Solids)

Explore multi-dimensional geometry and the sub-atomic realm... Jain will help you grasp the concepts of a Fourth Dimension and Time Travel.

Discover the importance of the Star Tetrahedron's 24 faces and 24 edges, and how this links to the 24 edges of the Cube Octahedron...the compressible and alchemical shape revered by Buckminster Fuller. Glimpse hidden mysteries within the 5 Platonic Solids and the 13 Archimedean Solids. Learn the 3-Dimensional form of the 5-pointed star that Pythagoras called the fifth element - Spirit aka Dodecahedron.

Learn the true value of Pi, based on Fractal Compression: Jain Pi = 3.14460...

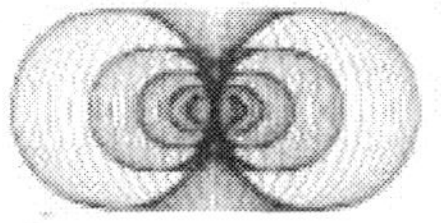

Divine Proportion (Lost Secrets of the 108 Phi Code)

Explore ancient hidden mysteries with Jain... why the shape of the pine cone is optically similar to the human heart, why we are awestruck when we view Sacred Architecture like the Parthenon of Greece and the Pyramids of Giza, and why we are attracted to the famous Mona Lisa.

Discover the secret codes concealed in the Gayatri Mantra, the most famous Eastern prayer for Enlightenment.

The human body is in resonance with the Living Mathematics of Nature. Embrace part of the Ascension process already morphologically and geometrically encoded into our bio-magnetic and bio-electrical fields.

For more information, contact:
jain@jainmathemagics.com www.jainmathemagics.com

(The original presentation of the 4 distinct topics, now changed to 5. Refer to the next 2 pages).

JAIN 108 MATHEMAGICS
SACRED GEOMETRY MYSTERY SCHOOL

- Level 1. Beginners Course — 5 Days of Remembering
- Level 2. Advanced Course — 5 Days of Awakening
- Level 3. Teacher Training — 5 Days of Mastery

ART of NUMBER

Level 1: • Digital Compression of the Multiplication Table revealing Atomic Structure of Rutile + Platinum Crystal

Level 2: • Prime Numbers 24 Pattern + 4th Dim. Templar Cross
• Binary Code — origin of VW Symbol

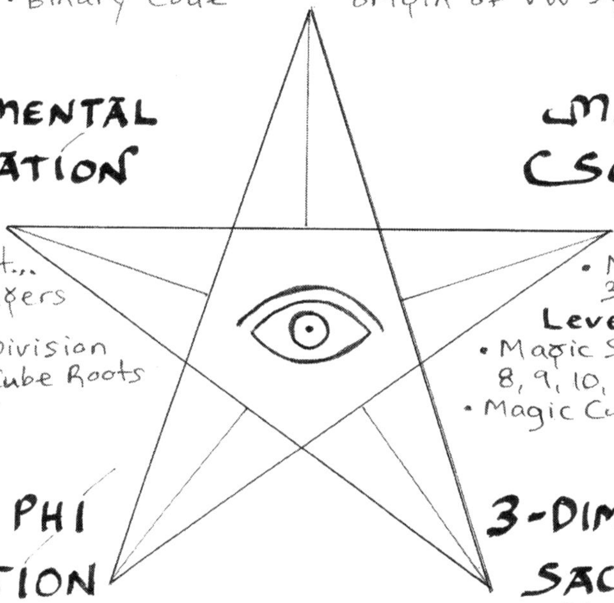

RAPID MENTAL CALCULATION

Level 1:
- Multiplicat...
- Magic Fingers

Level 2:
- One Line Division
- Square + Cube Roots
- Fractions

MAGIC SQUARES

Level 1:
- Magic Sqs. of 3, 4, 5, 6, 7

Level 2:
- Magic Squares of 8, 9, 10, 11, 12, 16
- Magic Cubes + Stars

DIVINE PHI PROPORTION (108 CODES)

Level 1: • Phi Spiral
- Fibonacci Numbers
- 108 Phi Code 1

Level 2: • Pentagram Constr.
- 108 Phi Code 2
- Mystical Squaring of Circle
- Vesica Pisces (the Mother of all Form)

3-DIMENSIONAL SACRED GEOMETRY

Level 1:
- the 5 Platonic Solids
- Fold Up Net of Dodecahedron

Level 2:
- the 13 Archimedean Solids
- Cuboctahedron perfected
- Truncated Icosahedron (Soccer Ball Shape).

JAIN F.R.E.E.D.O.M.S Non Profit Org...
trading as: JAIN MATHEMAGICS
www.jainmathemagics.com

(The original presentation of the 4 distinct topics, now changed to 5.
The new arm or addition is the very important concept of THE ART OF NUMBER which is typically now known as the 17th Sutra, it is what I am gifting to the world, by Jain 108 Principal of the EARTHHEART SCHOOL for Higher Learning).

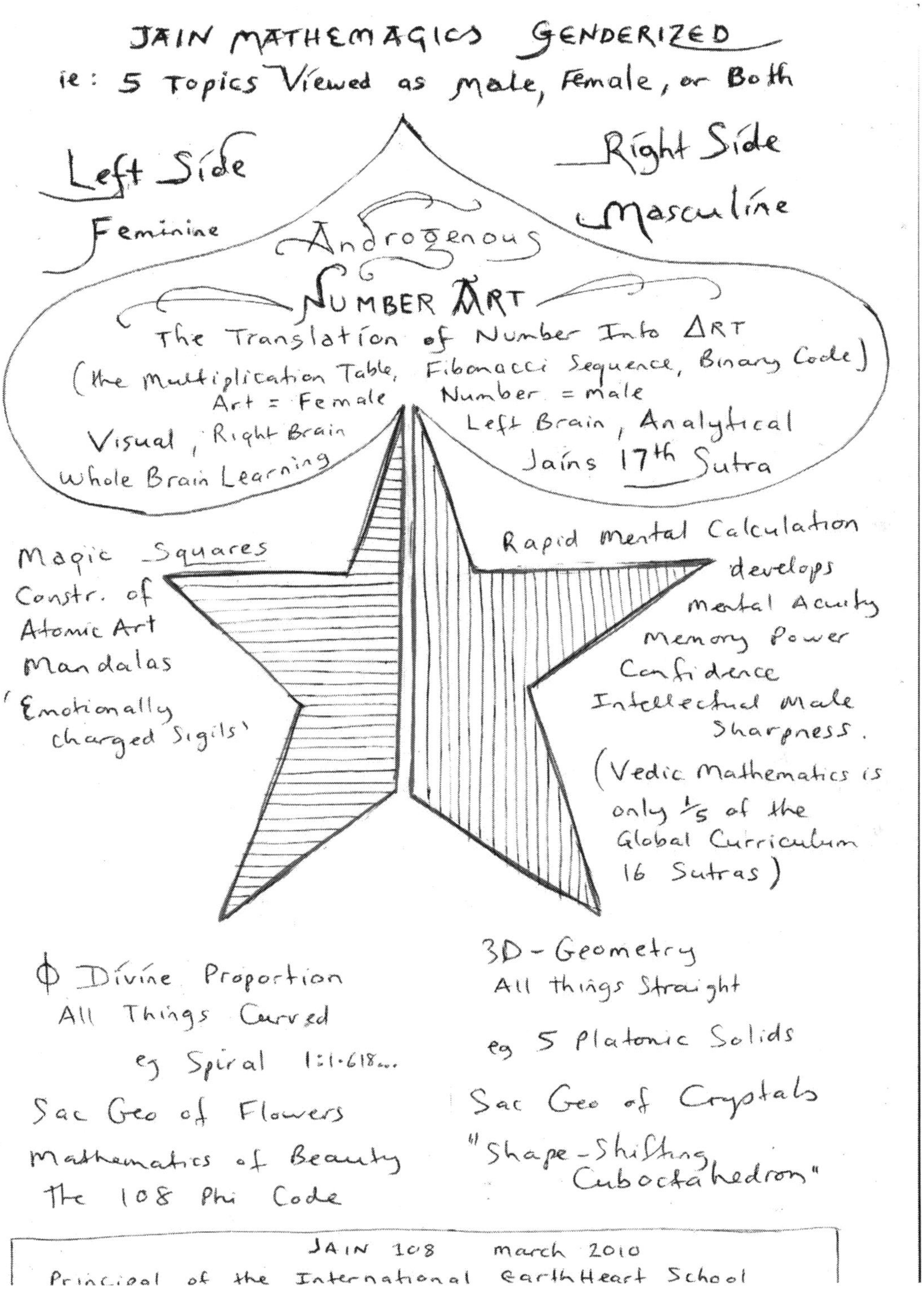

(The original presentation of the 4 distinct topics, now changed to 5.
The new arm or addition is the very important concept of THE ART OF NUMBER
which is typically now known as the 17th Sutra, it is what I am gifting to the world.
It is shown here with the 5 topics arranged into a genderized form
illuminating male and female qualities
by Jain 108 Principal of the EARTHHEART SCHOOL for Higher Learning).

OTHER BOOKs & DVDs By JAIN 108
On RAPID MENTAL CALCULATION
(aka Vedic Mathematics)

#1
THE MAGIC OF NINE IN VEDIC MATHEMATICS

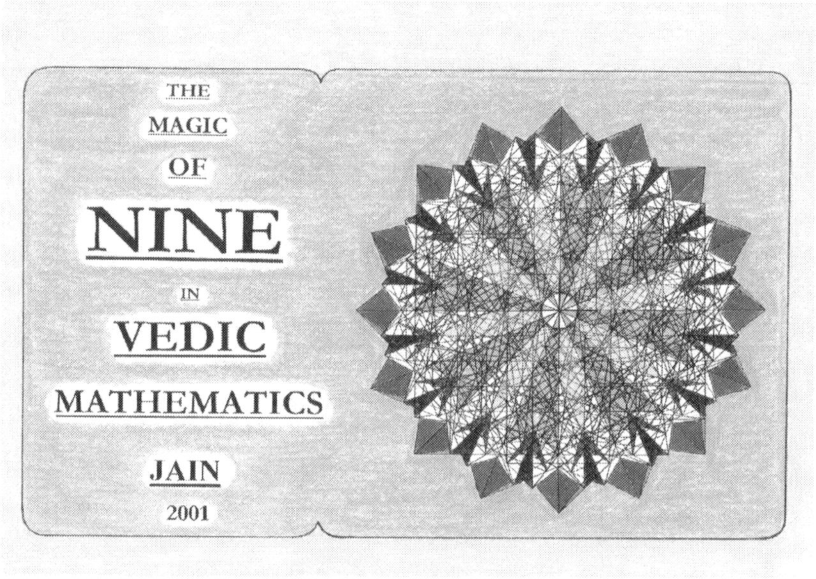

ISBN: 0-9594180-7-5, RRP = AUD $50, pub 2001, 124 pages, gms 390, size A4. Sun Yellow cover, landscape spiral bound. Abbreviated Code Name = M9

A highly intelligent and sophisticated system of Intuitive or Clairvoyant Mathematics.
This book is inspired by Shankacharaya of Puri a.k.a. Bharati Krsna Tirthaji's (1884-1960) now famous book 'Vedic Mathematics' which has spawned a 1,000 websites!
Shankacharaya's chief and current disciple is Maharishi Mahesh Yogi of Transcendental Meditation fame whose spiritual school is churning out child geniuses like the 'Rain Man'.
This book has a list of the 16 Vedic Sutras and explains some of them with examples.
It was originally compiled due to many requests after lectures to have a memory of Jain's Vedic Mathematics examples that were shown or demonstrated on the whiteboard.

#2

THE VEDIC MATHEMATICS CURRICULUM FOR THE GLOBAL SCHOOL.

Book 1: Digital Sums

ISBN:0-9594180-8-3, RRP = AUD $50, pub 2001, 142 pages, gms 430, size A4. Orange cover, landscape spiral bound.
Abbreviated Code Name = CURR1

(Part 1 of a 4 Part Series. Part 2 on MULTIPLICATION has been self-published in 2006).
The Series name has been changed to: JAIN MATHEMAGICS CURRICULUM FOR THE GLOBAL SCHOOL. Subtitled: Developing the Inner Mental Screen).
This book is everything possible on manipulating multi digits to single digits. This is the essence of 1 of 16 Vedic Sutras that means 'Digital Sums' or Digital Compression.
There is a whole chapter on Multiplication by 11 which is set out with questions and answers.
This book was the result of a 6 months Teacher Training course held in Byron Bay, and this book was an attempt to write the new material in a format with questions and answers.
This is the essence of Vedic Mathematics that it can be done mentally, via pattern recognition, or it can be done by one single line of intelligent arithmetic.

#3

JAIN MATHEMAGICS CURRICULUM

For The GLOBAL SCHOOL.

Book 2: MULTIPLICATION

Subtitled: Developing the Inner Mental Screen.

JAIN MATHEMAGICS CURRICULUM
for the GLOBAL SCHOOL

formerly known as: The VEDIC MATHEMATICS CURRICULUM for the GLOBAL SCHOOL

Part 2 MULTIPLICATION JAIN 2005

Developing the Inner Mental Screen

ISBN: 0-9757484-1-6, RRP = AUD $50, pub 2005, 160 pages, gms 500, size A4, Ochre Orange Cover, landscape spiral bound. Abbreviated Code Name = Bk2MULT

This is the book that most people request for at the end of a Jain Lecture or Seminar, something available for their children to commence applying practical, day-to-day examples.
It covers single digits x single digits 7 x 8 using ingenious Finger Computation, 2 digits x 1 digits, like 34 x 6, and 2 digits x 2 digits, like 97 x 97.

This is where the Teen Numbers For Teenagers gets introduced, inviting the parent to assist their child to know every Teen multiplication, like 14 x 15 or 16 x 18, within seconds of rapid mental calculation.

Lots of Examples, with lots of Exercises and all the Answers shown in full detail.

This book has long been waited for. There are special sections for Advanced Students and Teacher's Resources which include pages on Worksheets.
Hidden in the Answers Section is a sampling of Jain's Dictionary of Numbers aka **Harmonic Stairway** which illuminates dozens of anointed numbers revered by ancient Number enthusiasts.
It has an appendix with 2 superb published articles on Vedic Mathemagics that went global and launched Jain's career onto the international stage.

#4

JAIN MATHEMAGICS CURRICULUM
For The GLOBAL SCHOOL.
Book 3: MULTIPLICATION
Subtitled: Mental Calculations At The Speed Of Thought

JAIN MATHEMAGICS CURRICULUM for the GLOBAL SCHOOL

formerly known as:
The VEDIC MATHEMATICS CURRICULUM for the GLOBAL SCHOOL
(year 2006)

Book 3
MULTIPLICATION
@ The Speed Of Thought

JAIN 108
year 2011

This book is currently being written and will be released late 2011

ISBN: **978-0-9757484-6-6**

#5

RAPID MENTAL CALCULATION For TEENS And ADULTS WORKBOOK Subtitled: LEVEL ONE (aka Vedic Mathematics) by: JAIN 108

RAPID MENTAL CALCULATION
for TEENS and ADULTS
WORKBOOK

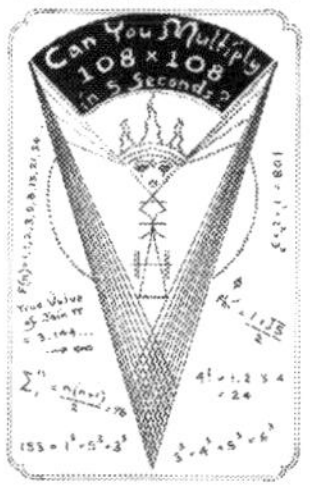

LEVEL ONE

From the series:
"JAIN MATHEMAGICS CURRICULUM FOR THE GLOBAL SCHOOL"

2011
(Script Material for the Film and OnLine Tutorials).

By **JAIN 108**
(writer, artist and Mathematical Futurist)

Produced by JAIN F.R.E.E.D.O.M.S
(For Research Expressing Essential Data Of Magic Squares)

ISBN: 978-0-9757484-5-9,
RRP = AUD $50, pub 2011,
175 pages, gms 500,
size A4, Yellow Colour Front & Back Cover,
Portrait spiral bound.
Abbreviated Code Name = Lev1Cur4

This booklet contains a well-thought out procedure for teaching the basics for Rapid Mental Calculation. It builds upon the 2 main foundations of knowing your Times Table and your Complements of 9 and 10.
This leads to the ability to confidently know the multiplication of your Teen Numbers like 13x14 or 12x18 etc. By knowing how to do this mentally and as a One-Line Answer, it is guaranteed that the student can manage any problem given to them over the course of 10 weeks which is a traditional Term or season of the School Year.
The format of the booklet is designed so that worksheets are printed on the Right Hand Side of the book, and the Left Hand Side is kept as blank, so that the student can do their working out on that side. It keeps all the material together; and for teachers, it assists in knowing how far the student has studied with his or her work.
This book is original in its style of content given and its delivery. It succeeds in giving capsules of easy to understand sutras or formula and deserves its place as a coherent and well articulated Global School Mathematics Curriculum.

CAN You MULTIPLY 108 x 108 in 5 Seconds?

RAPID MENTAL CALCULATION
For TEENS And ADULTS WORKBOOK
Subtitled: LEVEL ONE (aka Vedic Mathematics)
by: JAIN 108 of Australia
from the Series: JAIN MATHEMAGICS CURRICULUM For The GLOBAL SCHOOL.
Book 4: Workbook Level 1

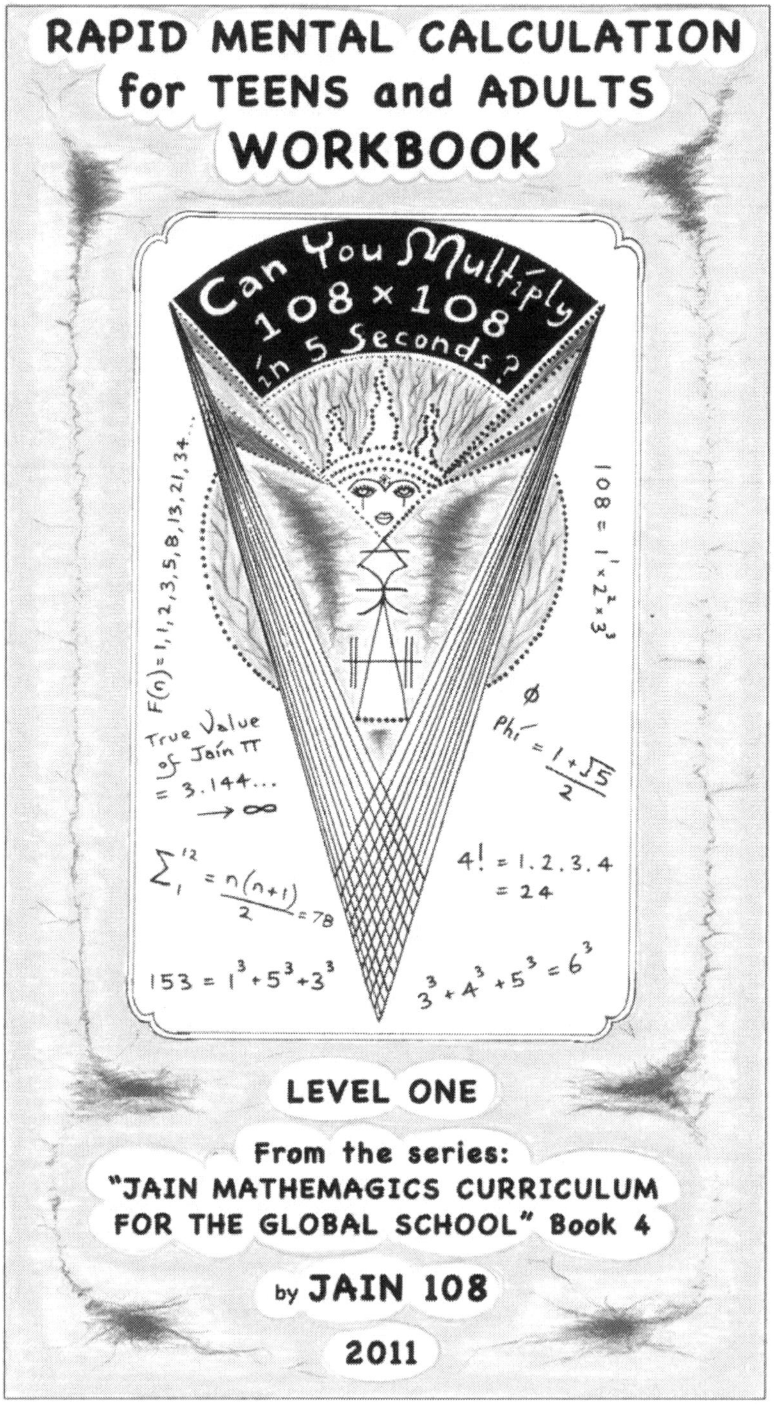

If you are allergic to Mathematics, then this series of books is for you. Goodbye to Factory Style Mathematics, the Nemesis of several generations of wounded students forced to be spoon-fed a non-organic, pasteurized and homogenized mathematical curriculum, inherited and tampered with since Euclid's time. And welcome to this modern Mathematical Whole Foods, a Bullet Proof system, serving you the choicest Mathematical Plums.

ABOUT The NEXT WORKBOOK: LEVEL 2

#6
RAPID MENTAL CALCULATION For TEENS And ADULTS WORKBOOK
Subtitled: LEVEL TWO (aka Vedic Mathematics)
by: JAIN 108

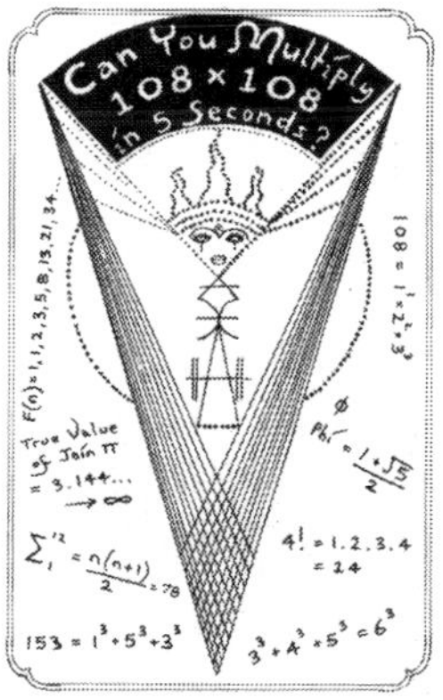

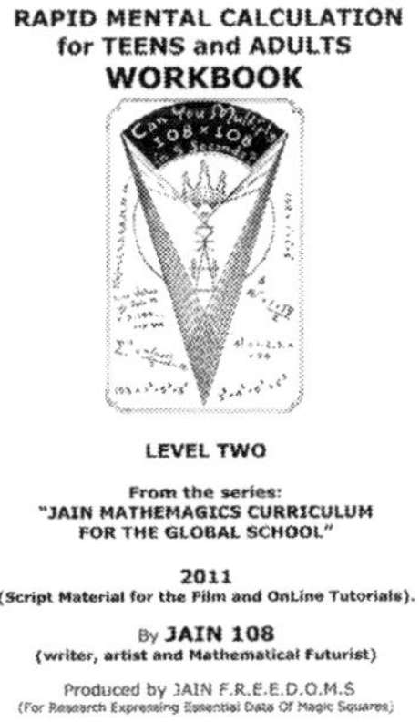

ISBN: 978-0-9757484-7-3, RRP = AUD $45, pub 2011, 177 pages, gms 500, size A4, Orange Colour Front & Back Cover, Portrait spiral bound.
Abbreviated Code Name = Lev2Cur4

Same style as the workbook above.

ABOUT The NEXT WORKBOOK: LEVEL 2
Some topics covered include:

~ **Pascal's Triangle** and its hidden patterns,
~ **Vertically and Crosswise** for multiplying 3Dx2D like 123x45 and 4Dx4D like 1234x5678,
~ **Gelosia** or Arabic Multiplication,
~ **Cubic Roots**. Be amazed how you can determine the Cube Root of a large number under 1,000,000 in about 2 seconds!
Eg: What is the Cube Root of: 35,937?
The Answer is 33. ie: 33x33x33 = 35,937.
No mathematician using the western techniques can do this, yet it is done quickly and simply by a teenager who understands the Universal Language of Pattern Recognition.
~ **Fractions**
~ **Percentages**

(colour front cover)

CAN You MULTIPLY 108 x 108 in 5 Seconds?

If you are allergic to Mathematics, then this series of quinta essential books is for you. ***Goodbye to Factory Style Mathematics***, the nemesis of several generations of wounded students forced to be spoon-fed a non-organic, stale, pasteurized and homogenized mathematical curriculum, inherited and tampered with since Euclid's time. And welcome to this ***modern Mathematical Whole Foods***, a well-researched and Bullet Proof system, serving you the choicest Mathematical Plums.

This Workbook continues on from where Level 1 ended. It goes beyond the boundaries of Vedic Mathematics which is the basics for Rapid Mental Calculation. It is ***guaranteed that the student can manage any problem*** given to him or her over the course of 10 weeks which is a traditional Term or season of the School Year.

This book is original in its coherent style of content and ***highly rich in psycho-active graphics***. It succeeds in giving capsules of easy to understand sutras or formula and is the beginning of a well articulated Global School Mathematics Curriculum.

Some topics covered include:

- ***Pascal's triangle*** and 5 Hidden Sequences within this Triangular Matrix like the Binary Code, the ***Fibonacci Sequence*** and the Powers of 11.
- Percentages and Fractions eg: Convert the fraction 7/40 as a Percentage.
- Squaring, ***Square Roots and Cube Roots*** eg: $\sqrt{4906}$ and $3\sqrt{35,937}$.
- ***The Magic Of 9***, & Extra Puzzles on Combinatorics and Probability.
- Sutra: Vertically & Crosswise for 3 Digits x 3 Digits and above. eg: 123 x 45.
- Squaring of Numbers Ending in 25 eg: 6252.
- ***Pythagoras' Theorem*** to determine the Roots of 2, Root 3 and Root 5.
- Revision Test called 'Mathletics'.

Jain 108

Students receive a Certificate of Completion.

JAINMATHEMAGICS.COM

(colour back cover)

#7

VEDIC MATHEMATICS
by: Bharati Krsna Tirthaji (1884-1960)

SRI SHANKARACHARYA AT SRF-YSS HEADQUARTERS

Sri Jagadguru Shankaracharya Bharati Krishna Tirtha of Puri, India, at Los Angeles international headquarters (founded in 1925 by Paramahansa Yogananda). In 1958 the senior head of the Swami Order paid a three-months' visit to America, sponsored by Self-Realization Fellowship. It was the first time in the history of the ancient Swami Order that a Shankaracharya had traveled to the West.

aka: The Shankacharaya of Puri. ISBN: 81-208-0164-4 published by Motilal Banarsidass. Paperback RRP AUD $45, has 358 pages, size A5, gms 460, Hardback RRP AUD AUD $55. same as above, gms 570, Abbreviated Code Name =VMBKT

A highly intelligent and sophisticated system of Intuitive or Clairvoyant Mathematics based on 16 Sutras or Word Formulae. Shankacharaya was invited to America by Paramahansa Yogananda's Self Realization Fellowship on a world tour in 1958 spreading the ancient message of Vedic Mathematics. Originally Bharati Krsna (Bharati means India) had written 16 volumes or 16 books, a book per Sutra. Unfortunately these books were confirmed lost by a disciple, whilst Bharati Krsna was in America and he was asked by his disciples to quickly write an Introduction to Vedic Mathematics, as people in mainland India were now imitating this material and claiming psychic powers and not acknowledging the Vedic source. On confirmation that all of the 16 books on the 16 Sutras were lost, and when he also had cataracts of the eyes and was going blind, Bharati Krsna's memory was so astute he was able to dictate the whole contents of this book from mere memory. It is chiefly an Introduction to the lost 16 books and implies that there is so much more to come. Currently India has set up a World Vedic Mathematics Academy and is offering a large reward for the retrieval of these now priceless lost 16 books (that we suspect is hidden in the Vatican!).

#8
CALCULATIONS @ The SPEED Of LIGHT
by **Kranti Kiran**

(pub:2002, 104pp) $45 230 gm

This book is like a Bible on Vedic Mathematics.
Kranti Tumma, age 29, currently lives and works in Hyderabad city, India.
Kranthi is a math wizard, who has made over 700 public performances including television and radio shows on calculations since February 1993.
At the age of 14 was a child prodigy and was referred to as a "walking computer". He has selflessly demonstrated the Magic of Pattern Recognition for half his life.
Seemingly complex calculations can be solved extraordinary swiftly. Even though he is considered "A Gift of God", his teachings are shareable and easy.
This book has jaw-dropping examples that are easily understood and replicable. There are many activities with answers in the back.
By going through this process, your Memory Power will Increase, and it will definitely boost your Mathematical Confidence.

What do I learn?

- You start off with refining skills in basic multiplication, basic division, basic addition, basic subtraction, etc. Then you learn:
- **Speed multiplication**: you learn many methods for **mentally** multiplying huge numbers: for example: 425974315 x 57493.
- **Squares**: learn to mentally square 2-digit numbers. Square 3-digit numbers by using little written working. **Cube**: learn to mentally square 2-digit numbers. Cube 3-digit numbers by using little written working.
- **Square roots**: calculate square roots of up to 4-digit numbers. **Cube roots**: calculate square roots of up to 9-digit numbers.
- **Calendar**: Can calculate the day of the week for any date.
- The best of all: you learn how to visualize all this in your mind. You develop powerful concentration. This is the key to all success.

DVDs on: VEDIC MATHEMATICS

#9
VEDIC MATHEMATICS FOR THE NEW MILLENNIUM (DVD)

ISBN : 0-9594180-9-1, RRP = AUD $45 for the dvd,
Recorded live 2002, 135 minutes, the DVD weighs 88 gms,
landscape spiral bound. Format: DVD PAL and NSTC available for USA.
Abbreviated Code Name = dvdVM

This is a highly acclaimed dvd, **being the first dvd in the world** teaching Rapid Mental Calculation. Jain, Mathematical Futurist, and EcoPreneur, is 10 years ahead of his time!
Richly illustrated with photos of the Masters and including relevant diagrams which makes learning easy, this program is the most popular of all Jain's products and is the best introduction to Vedic MatheMagics.
Find out how the 16 Sutras can solve every known mathematical problem. Currently, mathematicians have no known way of determining say the square root of 38. Jain shows how it can be effortlessly done to 6 decimal places in one line of working out.
Then, unbelievably, he mentally calculates the cube root of numbers in the hundreds of thousands!
You will never again multiply or divide the way you were taught at school.

#10
THE LIVING MATHEMATICS OF NATURE:
JAIN MATHEMAGICS
5 DVD SERIES:
2 of 5
VEDIC MATHEMATICS

ISBN ......, Length is 2 hours, gms 90. RRP = AUD $45, pub 2006, Has Phi Spiral and Magic Square Logo on Front Cover. Abbreviated Code Name = dvdLMoN2 or dvd 2/5

This is a studio produced dvd, with Jain explaining this unique and original material using a whiteboard, an overhead projector and geometrical models.
The order of the 38 sub-titles are:

Contents: Introducing Vedic Mathematics, Introducing Jain Mathemagics, Rote Learning, Dyslexic Children, Finger Computations, Universal Times Table, Merit for Maths, Human Bio-calculator, Digital Compression, Masters of Ancient Knowledge, Other Masters, Sutras / Formulas, Vertically and Cross-wise, The X Factor, From Flatland to Four Dimensions, Digital Compression *continued,* Teen Numbers for Teenagers, By The Excess, By One More Than The Previous Digit, Cognate Numbers, Pattern Recognition, Vertically and Cross-wise *continued,* Unity Consciousness, By The Deficiency, Ancient Shortcuts, Multiplication, Over The Base, Different Solutions, Using Another Sutra, Another Solution, Seeing The Answer, Base One Thousand, By Mere Observation, Mental One Line Division, Multi-Cultural Mathematics, Origin of Numbers, Training Manuals, Schools Teaching Vedic Maths, Jain's Background.

Discount: Save $25 when purchasing the whole set of 5 dvds on Jain Mathemagics. Reduced from $5 x $45 = $225, to $200.

#11

DVD: DOLPHIN HUMAN CONNECTION
Jain's 40 min Live Lecture on RAPID MENTAL CALCULATION (aka Vedic Mathematics)

RRP $55, a 40 minute lecture in this 2 and a half hours production, Abbreviated Code Name = dvdDHC

Jain was one of 4 key speakers at the Byron Bay Dolphin Therapy Day, May 2006, 40 min lecture, featuring presentations from Dolphin Therapy Day-(May 30th, 2006) talks and visual presentation and Sacred Sound Concert.

Presenters:
Jain-

Imagine if the children of the next generation could do all their Mathematics in their Head. Shape and Pattern Recognition will show how this Gift is connected to the Dolphin Mind.

www.jainmathemagics.com

Other Presenters: Andrew Parker, Liz Hawkins, Patricia Athena, Camilla, Elvina, Jennifer

Regarding Jain's Lecture:

It is based on a chronological presentation of about **40 original and exquisite colour drawings and murals** that Jain sketched and painted over a period of 25 years. It demonstrates how this **Dolphin Consciousness** has always been a deep part of his psyche and how these drawings were related to his innate geometrical understanding of the universe.

His lecture evolves to an understanding of a **Universal Language of Pattern Recognition** and then he demonstrates some of the sutras or formulae that govern Vedic Mathematics, the **supreme Art of Rapid Mental Calculation**. It concludes on an interesting note, how one of his students in a Cairns weekend seminar on Vedic Mathematics, who was semi-autistic, was able to perform mental calculations quicker than the other 30 participants of children and adults. When you ask these children how they can perform such fast mental calculation they invariably reply that they "see the answer". This relates directly to the **Feminine Right Brain** that understands pictures, shapes, holograms, sounds etc.... There appears to be growing confirmation that autistic children are able to more readily respond to our Cetacean relatives.

Jain succeeds and ties in the Human–Dolphin connection quite well in such a short time of 40 minutes. He analogizes that the dolphin can be listening to **8 radio stations simultaneously** and that these Beings are reflecting to us Humans that we also have this in-built capability to be using the totality of our Brains.

Jain will be also lecturing on the Mathematical Origin of Symbols, and using the "Symbol" or "Multi-Dimensional Language of Shape" as a higher form of inter-species Communication and Transmission.

To view some of the 40 original images that Jain projected, original drawings and paintings extending from 25 years ago, please click on the space where 5 images can be viewed. Jain has over 1,000 original drawings, and these will be compiled into Art and Poetry books.

(Poster image by Gaurav Tekriwal of India, Kolkata:
http://vedicmathsindia.blogspot.com/
inspired by my visit there in 2005 when I suggested he make a 5 dvd set on Vedic Mathematics, similar to my 5 dvd set, which he promptly achieved and started his business teaching).

With all of Jain's books and dvds, you will learn the basic principle of how to access your **INNER MENTAL SCREEN** and claim genius mind:

Can you multiply 108 x 108 in five seconds?
Learn **High Speed Rapid Mental Calculation** which will enable you to calculate much faster compared to the conventional systems which basically force feed your children to learn algebra and other non-interesting topics pushing your child to mentally shut down. As the child misses days at school and learns more material, the foundations are weak and soon the castle or mathematical tower of cards falls to the ground. The schools are doing it back-the-front, the wrong way; they must first teach the visual content of exciting galactic topics like magic squares, divine proportions, pascal's triangle, the 5 platonic solids etc, excite the children, then the child will open like a flower and imbue anything that we choose to feed them.

- ☐ **Eradicates** fear of Maths completely.
- ☐ Much **Improved Academic Performance** in School and Instant Results.
- ☐ **Sharpens** your mind, increases mental agility and intelligence.
- ☐ **Increases** your **speed** and accuracy. Rewires or reconfigures the neural pathways of the brain to function in their natural state.
- ☐ Get rid of your Electronic Calculator and Become a **Human Bio-Calculator** yourself.
- ☐ Improves **memory** and boosts self-confidence.
- ☐ **Cultivates** an Interest in Anointed Numbers.
- ☐ Develops your left and right sides of your brain promoting Whole Brain Learning.

ORIGINAL FRONT COVER FOR WORKBOOK LEVEL ONE

CAN YOU MULTIPLY 108 X 108 IN 5 SECONDS?

RAPID MENTAL CALCULATION for TEENS and ADULTS WORKBOOK

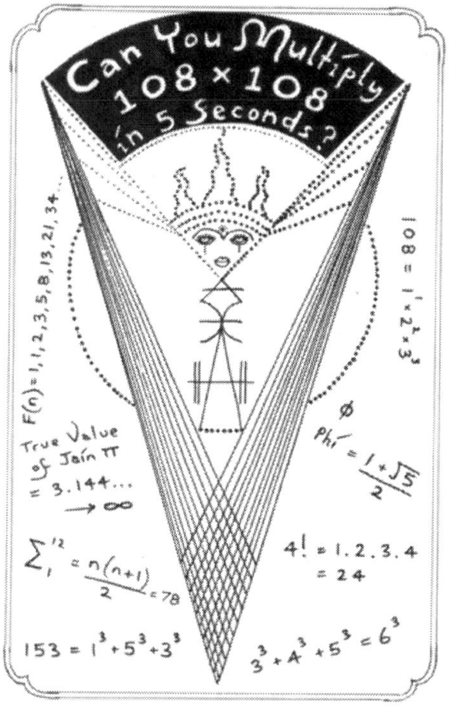

LEVEL ONE

From the series:
"JAIN MATHEMAGICS CURRICULUM FOR THE GLOBAL SCHOOL"

2011
(Script Material for the Film and OnLine Tutorials).

By **JAIN 108**
(writer, artist and Mathematical Futurist)

Produced by JAIN F.R.E.E.D.O.M.S
(For Research Expressing Essential Data Of Magic Squares)

Jain teaching the **108 Phi Code**, secret of the Divine Proportion.
The digitally compressed Fibonacci Sequence has an infinite repetition of 24 single digits based on 12 Pairs of 9, expressed as 108-9 forever, or it can be considered as a frequency of 117 the hertz or cycle per second relating to the King's Chamber in Gizeh.

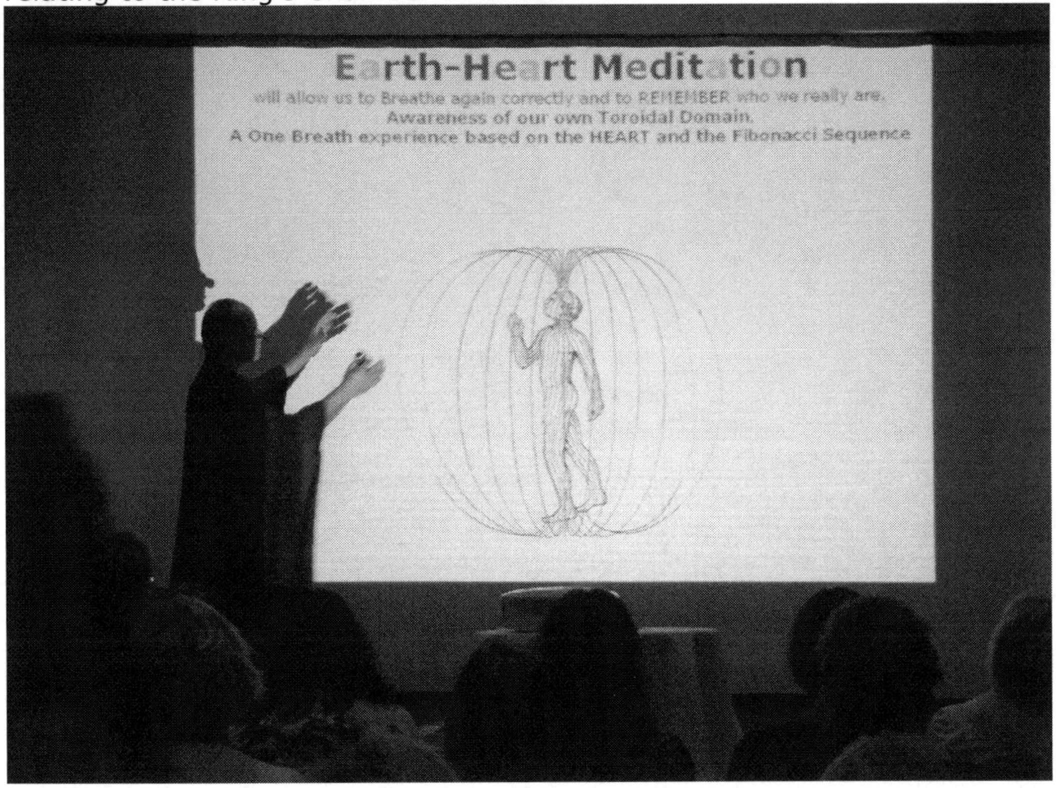

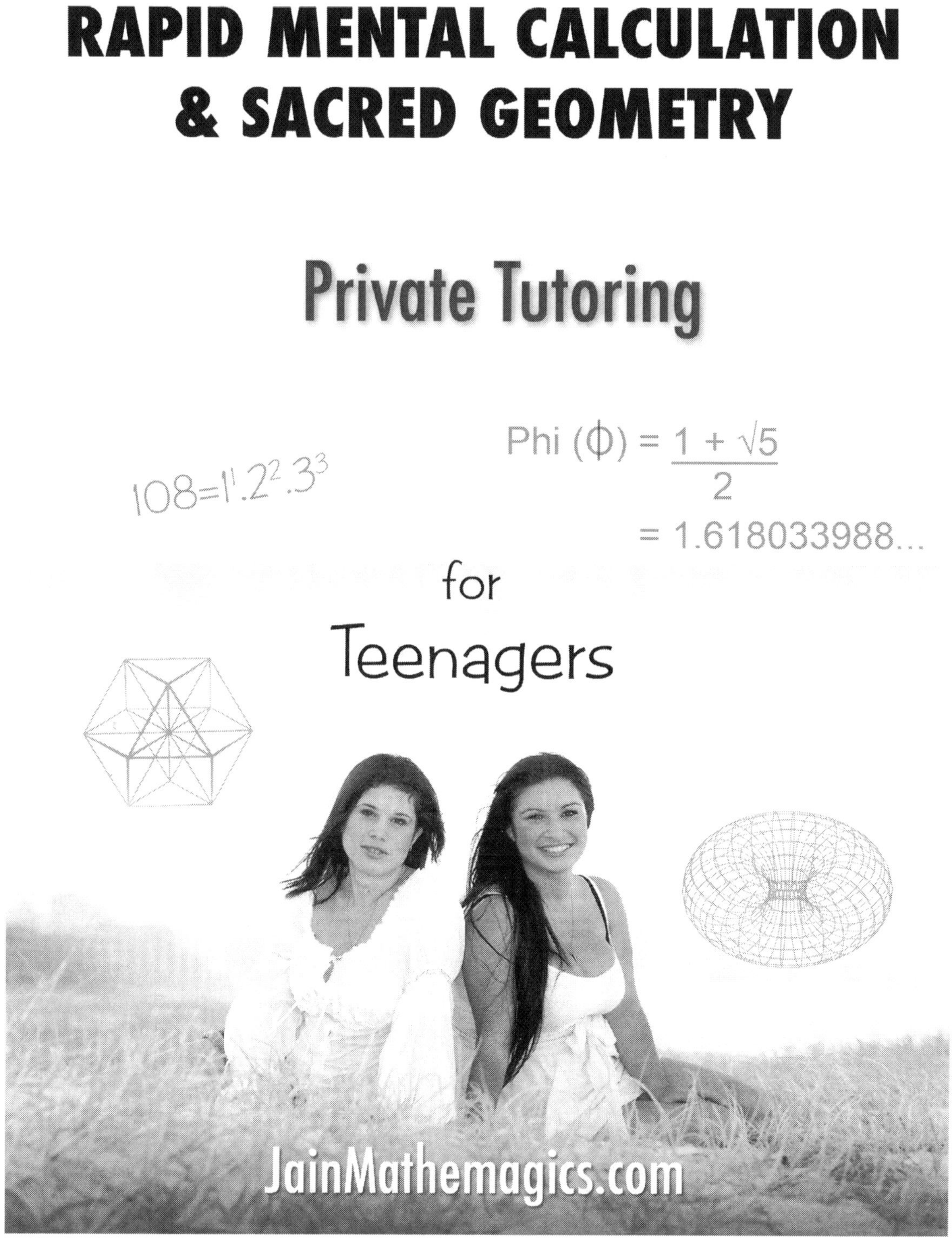

Poster advertising Private Mathematics Tutoring offered by Jain in the Mullumbimby, Byron Bay Shire, far north NSW, 2011. (Shown above: Mingkah Jain Sun on the left and Aysha Jain Sun).

INDEX

nb: all numbers in bold refer to that reference being an image.

C

D

M

N

O

P

Q

R

S

NUMBER INDEX

"Mathemagics For Juniors: Vedic Mathematics"

subtitle: Rapid Mental Calculation, Keen Teens and Adults

(front cover to Mathemagics For Juniors Workbook published 2015)

nb: This Workbook on Vedic Mathematics volume 5, (that you are reading now) can be considered a workbook for Adults, in contrast to another similar workbook on Vedic Mathematics: Rapid Mental Calculation, for young teenagers, called Mathemagic For Juniors.

If you enjoyed this 177 page book, and would like a simplified version with 111 pages, then you can order this book:

http://www.jainmathemagics.com/product/141/
To view and order the Ebook on line, click on:**http://www.jainmathemagics.com/product/151/**

Size: *A4 portrait* ***Cost***: *$40* ***ISBN: 978-0-9872543-7-5***
Self Published by JAIN 108 MATHEMAGICS in 2015 First Edition, Byron Bay

CAN YOU MULTIPLY 108 x 108 IN 5 SECONDS?

• This **9 Lesson / 9 Week Course** will set you on your path to claim your natural mathematical **Genius**, to build both your **Memory Power** and **Confidence**. **Become a Human Bio-Calculator**!
• This workbook is excellent for all public and private schools offered as supplementary or complementary curriculum lessons.
• It is designed to have the parent sitting in the class with their child. Ultimately, the aim of this book is to get the student to apply knowledge confidently in practical situations, like being in a supermarket, and mentally calculating the change from $10 when buying 9 apples at 69 cents each, before the girl at the check-out calculates it electronically.
• Jain 108 has been actively teaching **Vedic Mathematics**, and **Sacred Geometry** for the last 30 years, overseas and around Australia.
• He produced **the first dvd in the world** on **Vedic Mathematics For the New Millennium** in 2002, to cater for the growing online curriculum to fulfill his task of making this rare and ancient content available for the **international home-schooling movement**.
• Jain's focus for this university-style learning is towards **Juniors** (9-12 yo) and **Keen Teens**. This **Mathemagics** material is a veritable **Miniversity**, known as the **EarthHeart School** in Byron Bay and known as **The Joy Of Numbers Institute** in Asia. This same workbook is used to teach curious adults.
• As an adjunct to this book, there are **dvds** available based on live filming of these classes in Mullumbimby (Byron Bay area, far north NSW). The dvds show the true progression of a typical child not knowing these essential sutras or formula, then demonstrating to themselves that they were genuinely able to perform Rapid Mental Calculation, at the Speed of Thought.
• Tribute is given to **Bharati Krsna Tirthaji** who released these 16 sutras about 50 years ago.

VEDIC MATHEMATICS

WORKBOOK

- This 9 Lesson / 9 Week Course will set you on your path to claim your natural mathematical Genius, to build both your Memory Power and Confidence. Become a Human Bio-Calculator!
- This workbook is excellent for all public and private schools offered as supplementary or complementary curriculum lessons.
- It is designed to have the parent sitting in the class with their child. Ultimately, the aim of this book is to get the student to apply knowledge confidently in practical situations, like being in a supermarket, and mentally calculating the change from $10 when buying 9 apples at 69 cents each, before the girl at the check-out calculates it electronically.
- Jain 108 has been actively teaching Vedic Mathematics, and Sacred Geometry for the last 30 years, overseas and around Australia. He produced the first dvd in the world on Vedic Mathematics For the New Millennium in 2002, to cater for the growing online curriculum to fulfill his task of making this rare and ancient content available for the international home-schooling movement.
- Jain's focus for this university-style learning is towards Juniors (9-12 yo) and Keen Teens. This Mathemagics material is a veritable Miniversity, known as the EarthHeart School in Byron Bay and known as The Joy Of Numbers Institute in Asia. This same workbook is used to teach curious adults.
- As an adjunct to this book, there are dvds available.
- Tribute is given to Bharati Krsna Tirthaji who released these 16 sutras 50 years ago.
- Some of the Chapter Contents include:

- Magic Fingers to multiply single digits eg: 7x8, and 9x38 and the Moroccan Method to multiply 13x14
- Multiplication by 11 using the sutra: Digital Compression
- Sutra: All From 9 & The Last From 10 to subtract 3,144 from 10,000
- Sutra: By The Excess to solve teens being multiplied like 12x18 or 108x108 or 1,011x1,034
- Sutra: By The Deficiency to solve 98x87
- Trachtenberg Method to single digits by two digits eg 4x24 or 6x6x6
- Sutra: By One More for the Squaring Of Numbers Ending In 5 eg: 35x35 or 452
- Sutra: Vertically and Crosswise to multiply 2 digits by 2 digits like 23x78 and much more

CAN YOU MULTIPLY 108 x 108 IN 5 SECONDS?

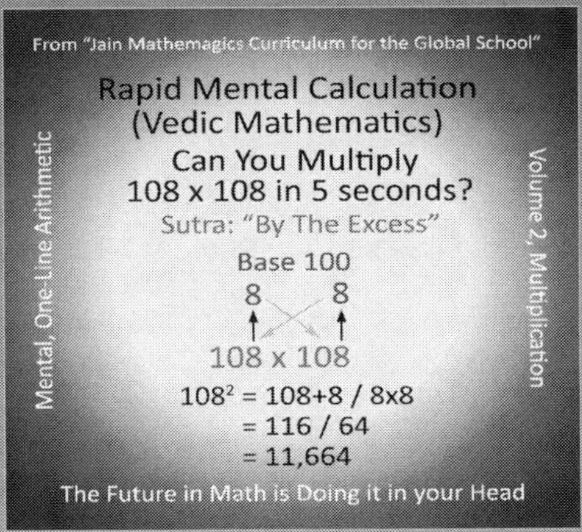

The Joy of Numbers Institute

jainmathemagics.com

PUBLISHED 2015

(back cover to Mathemagics For Juniors Workbook published 2015)

The End

Made in United States
Orlando, FL
18 November 2025

72700551R00098